Science & Life Issues

MY BODY AND ME

SCIENCE EDUCATION FOR PUBLIC UNDERSTANDING PROGRAM
SEPUP

UNIVERSITY OF CALIFORNIA AT BERKELEY
LAWRENCE HALL OF SCIENCE LHS

LAB-AIDS
RONKONKOMA, NEW YORK

This book is part of SEPUP's middle school science course sequence:

Issues and Earth Science

Rocks, Minerals, and Soils
Shaping the Land
Weather and Atmosphere
Earth and the Solar System

Science and Life Issues

My Body and Me
Micro-Life
Our Genes, Our Selves
Ecology and Evolution
Using Tools and Ideas

Issues and Physical Science

Properties of Materials
Water
Energy
Force and Motion

Additional SEPUP instructional materials include:

CHEM-2 (Chemicals, Health, Environment and Me): Grades 4–6
SEPUP Modules: Grades 7–12
Science and Sustainability: Course for Grades 9–12

This project was supported, in part, by the
National Science Foundation
Opinions expressed are those of the authors and not necessarily those of the Foundation.

2 3 4 5 6 7 8 9   09 08 07 06 05

ISBN: 1-887725-30-X

For photo and illustration credits, see page B-95, which constitutes an extension of this copyright page.

The preferred citation format for this book is:
SEPUP. (2005). *My Body and Me* (Units A and B of *Science and Life Issues.*) Lawrence Hall of Science, University of California at Berkeley. Published by Lab-Aids®, Inc., Ronkonkoma, NY.

SEPUP
Lawrence Hall of Science
University of California at Berkeley
Berkeley CA 94720-5200
e-mail: sepup@berkeley.edu
Website: www.sepuplhs.org

Published by:
LAB-AIDS
17 Colt Court
Ronkonkoma NY 11779
Website: www.lab-aids.com

A Letter to SALI Students

As you examine the activities in this book, you may wonder, "Why does this book look so different from other science books I've seen?" The reason is simple: it is a different kind of science program, and only some of what you will learn can be seen by leafing through this book!

Science and Life Issues, or *SALI,* uses several kinds of activities to teach science. For example, you will design and conduct an experiment to investigate human responses. You will explore a model of how species compete for food. And you will play the roles of scientists learning about the causes of infectious disease. A combination of experiments, readings, models, debates, role plays, and projects will help you uncover the nature of science and the relevance of science to your interests.

You will find that important scientific ideas come up again and again in different activities. You will be expected to do more than just memorize these concepts: you will be asked to explain and apply them. In particular, you will improve your decision-making skills, using evidence and weighing outcomes to decide what you think should be done about scientific issues facing society.

How do we know that this is a good way for you to learn? In general, research on science education supports it. In particular, the activities in this book were tested by hundreds of students and their teachers, and they were modified on the basis of their feedback. In a sense, this entire book is the result of an investigation: we had people test our ideas, we interpreted the results, and we revised our ideas! We believe the result will show you that learning more about science is important, enjoyable, and relevant to your life.

SALI Staff

SEPUP STAFF

Dr. Barbara Nagle, Director
Manisha Hariani, Associate Director
Lee Amosslee, Instructional Materials Developer
Janet Bellantoni, Instructional Materials Developer
Asher Davison, Instructional Materials Developer
Kate Haber, Instructional Materials Developer
Daniel Seaver, Instructional Materials Developer
Miriam Shein, Publications Coordinator
Roberta Smith, Administrative Coordinator
Ezequiel Gonzalez, Administrative Assistant

Dr. Herbert D. Thier, Founding Director

CONTRIBUTORS/DEVELOPERS

Barbara Nagle
Manisha Hariani
Herbert D. Thier
Asher Davison
Susan K. Boudreau
Daniel Seaver
Laura Baumgartner

TEACHER CONTRIBUTORS

Kathaleen Burke
Richard Duquin
Donna Markey

CONTENT AND SCIENTIFIC REVIEW

Gary R. Cutter, Director of Biostatistics, AMC Cancer Research, Denver, Colorado

Peter J. Kelly, Emeritus Professor of Education and Senior Visiting Fellow, School of Education, University of Southampton, Southampton, England

Deborah Penry, Assistant Professor, Department of Integrative Biology, University of California at Berkeley, Berkeley, California

RESEARCH ASSISTANCE

Marcelle Siegel, Leif Asper

PRODUCTION

Project coordination: Miriam Shein
Production and composition: Seventeenth Street Studios
Cover concept: Maryann Ohki
Photo research and permissions: Sylvia Parisotto
Editing: WordWise

Field Test Centers

The classroom is SEPUP's laboratory for development. We are extremely appreciative of the following center directors and teachers who taught the program during the 1998–99 and 1999–2000 school years. These teachers and their students contributed significantly to improving the course.

REGIONAL CENTER, SOUTHERN CALIFORNIA
Donna Markey, *Center Director*
Kim Blumeyer, Helen Copeland, Pat McLoughlin, Donna Markey, Philip Poniktera, Samantha Swann, Miles Vandegrift

REGIONAL CENTER, IOWA
Dr. Robert Yager and Jeanne Bancroft, *Center Directors*
Rebecca Andresen, Lore Baur, Dan Dvorak, Dan Hill, Mark Kluber, Amy Lauer, Lisa Martin, Stephanie Phillips

REGIONAL CENTER, WESTERN NEW YORK
Dr. Robert Horvat, *Center Director*
Kathaleen Burke, Dick Duquin, Eleanor Falsone, Lillian Gondree, Jason Mayle, James Morgan, Valerie Tundo

JEFFERSON COUNTY, KENTUCKY
Pamela Boykin, *Center Director*
Charlotte Brown, Tara Endris, Sharon Kremer, Karen Niemann, Susan Stinebruner, Joan Thieman

LIVERMORE, CALIFORNIA
Scott Vernoy, *Center Director*
Rick Boster, Ann Ewing, Kathy Gabel, Sharon Schmidt, Denia Segrest, Bruce Wolfe

QUEENS, NEW YORK
Pam Wasserman, *Center Director*
Gina Clemente, Cheryl Dodes, Karen Horowitz, Tricia Hutter, Jean Rogers, Mark Schmucker, Christine Wilk

TUCSON, ARIZONA
Jonathan Becker, *Center Director*
Peggy Herron, Debbie Hobbs, Carol Newhouse, Nancy Webster

INDEPENDENT
Berkeley, California: Robyn McArdle
Fresno, California: Al Brofman
Orinda, California: Sue Boudreau, Janine Orr, Karen Snelson
Tucson, Arizona: Patricia Cadigan, Kevin Finegan

Contents

UNIT A Studying People Scientifically

UNIT B Body Works

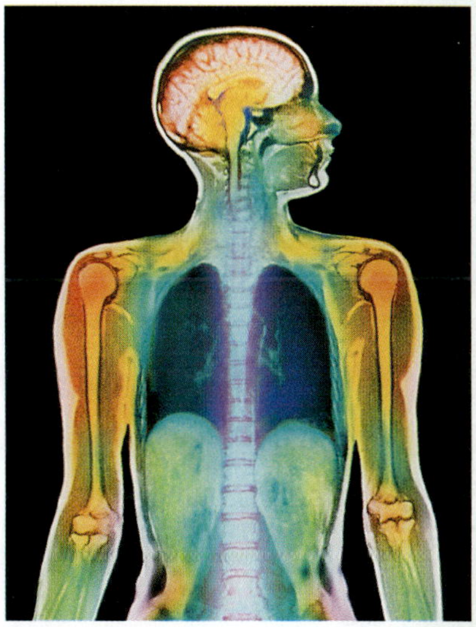

Studying People Scientifically

A

Unit A

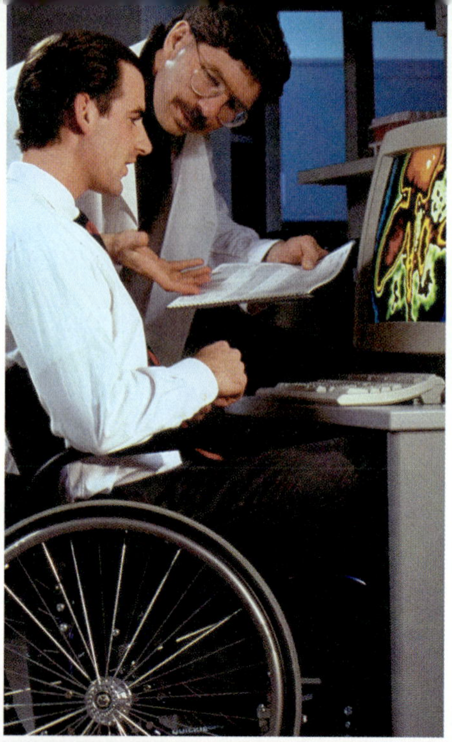

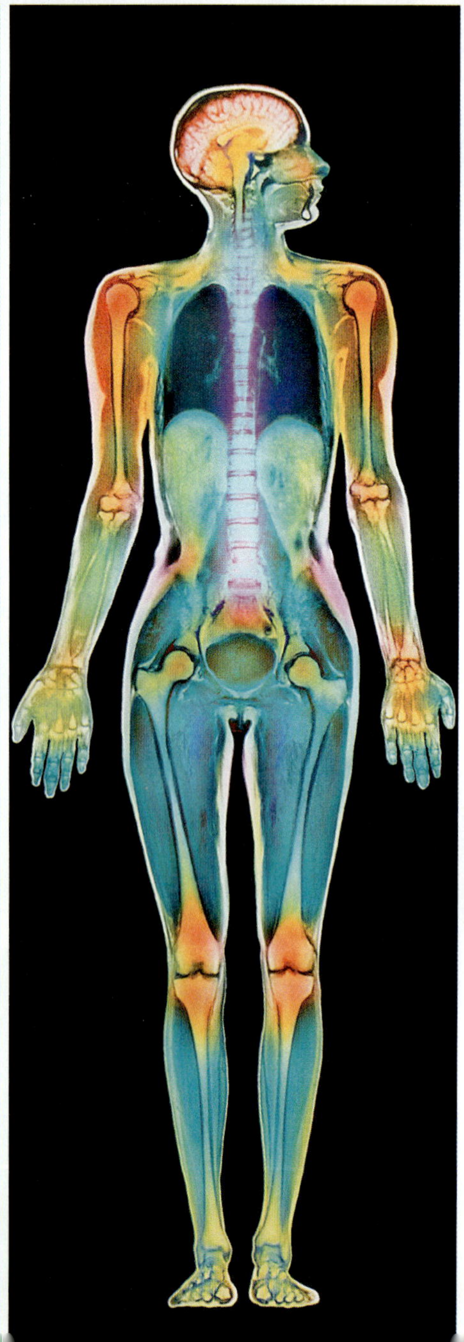

MEDICAL RESEARCH
STUDY PARTICIPANTS NEEDED
Be part of a medical research study and earn up to
a day. Your participation could help improve the
of life for people with various health
YOU CAN MAKE A DIFFERENCE! Study
who meet the following criteria are eligible:
18 to 68 years old
male or female

Studying People Scientifically

"But it's true. I heard about it on the news last night. Broccoli decreases your risk of having cancer," stated Alissa.

Eric rolled his eyes. "They used to say margarine was much better for you than butter, then it was butter is better—who knows what medical researchers are saying today?"

"I don't know, Eric, this sounded pretty convincing," replied Alissa. "They've done scientific studies and everything."

"I'm not going to eat it just because some doctor says I should," said Eric stubbornly.

• • •

What would you do if you were Eric? What factors might influence your view? How do you respond to everyday dilemmas, and on what basis do you make your decisions?

Scientists collect information and do studies to answer such questions. In this unit, you will learn some of the scientific principles and approaches used to study people.

1 Solving Problems: Save Fred!

INVESTIGATION

We all solve problems every day—from little problems like what to wear to school, to bigger problems like whether to get a job or go to college. Different kinds of problems require different problem-solving approaches. One problem can often be solved in more than one way. What do you do when you are faced with a problem?

CHALLENGE

How can you and your partner solve a problem?

MATERIALS

For each pair of students

1 **gummy candy life preserver**
1 **gummy candy worm**
1 **plastic cup**
4 **paper clips**

PROCEDURE

1. Work with your partner to solve the problem below. A picture of the set-up is on the next page.

SAVE FRED!

Poor Fred! He was sailing along on a boat (your plastic cup) when a strong wind blew it upside-down. Fred (your candy worm) ended up on top of the upside-down boat. Unfortunately for Fred, his life preserver (your candy life preserver) is still trapped under the boat.

Your job is to place the life preserver firmly around Fred's body, but you must obey three rules:

1. Fred may not fall into the "sea" (onto the table) more than one time; if he does, Fred "drowns."

2. You may not injure him in any way.

3. You may use only the four paper clips to move Fred, the boat, and the life preserver. You may not touch anything except the paper clips.

2. Work with your partner to record in your science notebook exactly what you did to save Fred. You may wish to draw a picture or a diagram to explain your procedure.

3. Explain your procedure to another two partners.

ANALYSIS

Each activity in this book asks Analysis Questions. These questions will guide you in your learning. You will not have to write your response to all of the questions nor will you have to answer all of them by yourself. Use the key below as well as directions from your teacher to find out exactly what to do.

KEY TO ANALYSIS ICONS

 = Answer the question by yourself.

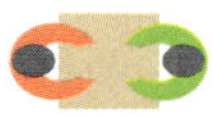 = Discuss with your partner.

 = Discuss with your group.

 = Discuss the question in class.

1. You can solve problems in many different ways. In fact, you may use more than one way to solve a single problem. You can

- develop a plan.
- find a pattern.
- draw a picture or a diagram.
- act out the problem.
- make a list.
- guess and test.
- work backward.
- write an equation.
- construct a table or graph.
- simplify the problem.
- use objects to model the problem.

Which of these ways did you and your partner use to save Fred?

2. As a class, discuss the ways in which various groups of partners saved Fred. How were your problem-solving methods similar? How were they different?

3. **Reflection:** People face problems in their lives every day. What did you learn from this activity that you can use to solve other problems?

2 The Pellagra Story

Scientists have common approaches to solving problems. Usually, people think of experiments as one way scientists investigate problems. What kinds of experiments are possible when you study human beings? How can you collect evidence in these situations? Begin to consider these issues as you watch the story of the disease called pellagra (puh-LAY-gra), which affected poor rural families of the South.

Poor families of the South were more likely to develop pellagra.

Whenever you see this icon, it means that you will find more information or a technology extension on this subject on the SALI page of the SEPUP website.

CHALLENGE

What are the common elements shared by all scientific problem-solving methods? How is science used to study people?

PROCEDURE

1. In order to prepare to watch the story on the video, first read Analysis Questions 1–4.
2. Your teacher will provide you with a student sheet or ask you to prepare a table like Table 1 on the next page to record your notes during the video.
3. Watch the story of pellagra on the video, *A Science Odyssey*: "Matters of Life and Death."

Table 1: Notes on The Pellagra Story

What was the problem of pellagra?

What did people think caused pellagra?

1.
2.
3.
4.

What evidence did Dr. Goldberger observe or collect about pellagra?

What did Dr. Goldberger conclude about the cause of pellagra?

ANALYSIS

Dr. Joseph Goldberger

1. **a.** What was the first step in Dr. Goldberger's research into pellagra? Explain why this step was important in developing his hypothesis.

   b. During this first step in his research, what evidence did Dr. Goldberger find that suggested that pellagra was not caused by germs?

2. **a.** What was Dr. Goldberger's hypothesis about the cause of pellagra?

   b. What did he do to provide evidence of the relationship between pellagra and nutrition? Be sure to explain *how* his research provided evidence that supported or disproved his hypothesis.

   c. How could he have provided more convincing evidence of the relationship between pellagra and nutrition?

3. Why didn't people believe Dr. Goldberger's conclusion about the cause of pellagra? Give two reasons.

4. Compare the steps of the traditional scientific method to the steps Dr. Goldberger followed to investigate pellagra. How were the steps the same? How were the steps different?

5. To investigate his hypothesis, Dr. Goldberger had prisoners volunteer to be fed a poor diet; as a result, 7 out of 11 prisoners developed pellagra. What do you think about Dr. Goldberger's decision to experiment on people? Support your answer with evidence and identify the trade-offs of your decision.

   Hint: To write a complete answer, first state your opinion. Provide two or more pieces of evidence that support your opinion. Then consider all sides of the issue and identify the trade-offs of your decision.

6. **Reflection:** How do people in different careers solve problems? Scientists, plumbers, engineers, auto mechanics, nurses, teachers, and many other workers solve problems. Choose two careers that interest you. Describe the kind of problems you think people face in these careers. Describe how you think they solve them.

3 Testing Medicines: A Clinical Trial

INVESTIGATION

Collecting evidence is an important part of science. One way to collect evidence is to conduct experiments as Dr. Goldberger did. Products such as medicines are tested by volunteers before they are made available to the public. These tests are known as **clinical trials**.

A CLINICAL TRIAL

Imagine that you suffer from severe headaches several times a month. These headaches are so painful that you can't read, listen to music, or watch television. Regular headache medicines don't work very well for you. One day, you complain to your doctor about your headaches. She tells you that the local medical school is conducting clinical trials of a new headache medicine. She asks if you would like to volunteer to be a part of this trial. Hoping for relief, you say yes.

Since medicines cannot be tested in the classroom, you will participate in a simulation of a clinical trial. In this simulation, differences in taste will equal differences in response to the medicine. Figure 1, "Clinical Trial of A Headache Medicine," on the next page, explains this simulation.

CHALLENGE

How are medicines tested during a clinical trial?

Figure 1: Clinical Trial of A Headache Medicine

The taste of the yellow lemon drink represents a headache.

The taste of the pink lemon drink represents the medicine taken to treat your headache.

If the pink lemon drink tastes the same as the yellow, there is no change in your headache.

If the pink lemon drink tastes better than the yellow, your headache is gone!

If the pink lemon drink tastes worse than the yellow, your headache is gone, but you experience side effects.

MATERIALS

For each group of four students

4 small tasting cups (one for each student)
1 sample cup of yellow lemon drink
1 sample cup of pink lemon drink

For each student

1 Student Sheet 3.1, "Analysis of Clinical Trial"

SAFETY

Never taste materials or eat or drink in science class unless specifically told to do so by your teacher. Be sure that your work area is clean and free of any materials not needed for this activity. If you are allergic to lemons or other citrus fruits, juice drinks, or sugar, or if you have any other health issue, such as diabetes, that limits what you can eat, tell your teacher and do not taste the drink samples in this activity.

PROCEDURE

1. Record your group number (found on the sample cups) in your science notebook; this represents the batch of medicine you received.
2. Fill your tasting cup half-full of yellow lemon drink by carefully pouring from the sample cup into your tasting cup.
3. Taste the yellow lemon drink. Empty the cup.
4. Fill your tasting cup half-full of pink lemon drink.
5. Taste the pink lemon drink.
6. *Did the pink lemon drink taste the same, better, or worse than the yellow lemon drink?* Record your response in a table like Table 1 on the next page.

Table 1: Results of Treatment

	Same as yellow lemon drink	Better than yellow lemon drink	Worse than yellow lemon drink
My response (Show with an X)			
My group's response (Show number of each)			

7. Share your results with your group. Summarize your group results in Row 2 of your data table.

8. Have one person from your group report your group's results to your teacher.

9. After a class discussion of the results, record the class's results and create a bar graph of the class's data on Student Sheet 3.1, "Analysis of Clinical Trial."

ANALYSIS

1. What evidence do you have that the medicine does or does not work to improve headaches?

2. **a.** What is a placebo?

   b. Why is a placebo group included in clinical trials?

3. In this activity, if a person finds that the drink tastes worse, the headache is gone, but there are side effects.

a. Assume that the side effects are mild, such as a slight stomachache. Explain why this medicine should or should not be sold to people suffering from a headache. Are there any trade-offs involved in your decision?

b. What if the side effects were serious, such as nausea and vomiting? Explain why this medicine should or should not be sold to people suffering from a headache. Are there any trade-offs involved in your decision?

4. In this activity, if a person finds that the medicine tastes better or worse, the headache is gone. Review the results of this simulation. Think about whether the medicine works and how often side effects occur. What would you conclude about the safety and effectiveness of this medicine for treating headaches? Support your conclusion with evidence.

EXTENSION

Go to the SALI page of the SEPUP website to find out how to post your class results on the site. Look at the results posted by other students. How do your results compare?

4 Testing Medicines Scientifically

How can we be certain that the foods we eat are safe, that the cosmetics we use won't harm us, and that medical products are effective? In the United States, the Federal Food, Drug, and Cosmetic Act and several other laws protect the public's safety. Before the Federal Food, Drug, and Cosmetic Act was passed in 1938, most products were not regulated.

CHALLENGE

How are medicines, such as over-the-counter and prescription drugs, tested?

READING

The Role of the FDA

Today, the U.S. Food and Drug Administration (FDA) enforces the laws on product safety and effectiveness. The FDA approves medicines that have been proven to be safe and effective for people to use. To be approved, a product must undergo and pass scientific tests. Usually, new medicines are first tested in animals to see whether they cause any harmful side effects. Only then are the medicines tested by volunteers in clinical trials. The results of clinical trials must show that the medicine is both safe and effective before it can be approved for use by the public.

STOPPING TO THINK 1

What are two reasons that medicines must be tested before they are made available to the public?

Informed Consent

In clinical trials, volunteers are chosen carefully. They must not be allergic to drugs similar to the one that will be tested. Also, they cannot be taking certain other medications. These things might make it more likely for them to experience harmful side effects. The volunteers must be told about any risks, such as possible side effects. They sign an **informed consent** form. This form states that they have been told (informed) about the risks and that they agree (consent) to participate in the trial.

STOPPING TO THINK 2

a. Why do volunteers in a clinical trial have to sign informed consent forms?

b. Make a list of all of the information that you think should be included in an informed consent form.

Treatment and Control Groups

The volunteers are then divided into two groups. One group receives a pill or liquid that contains the medicine being tested. The other group receives a pill or liquid that looks the same but contains only an inactive ingredient, such as sugar. This inactive pill is called a **placebo** (pla-SEE-bo). Neither the patients nor their doctors know who is receiving the medicine and who is receiving the placebo. The placebo is used as a **control**. It helps to prove that any improvement in the patient is due to the medicine and not to other aspects of the medical treatment. For example, just going to see a doctor may result in an improvement in a patient's condition.

STOPPING TO THINK 3

a. Explain the relationship between a placebo and a control.

b. Why do you think the doctor isn't told whether the patient is receiving the treatment or the placebo? Imagine you are the doctor. How would you act if you thought your patient was getting a treatment that will work? How would you act if you thought your patient was getting just a placebo? How might this affect the patient?

c. What is the purpose of a placebo?

The Placebo Effect

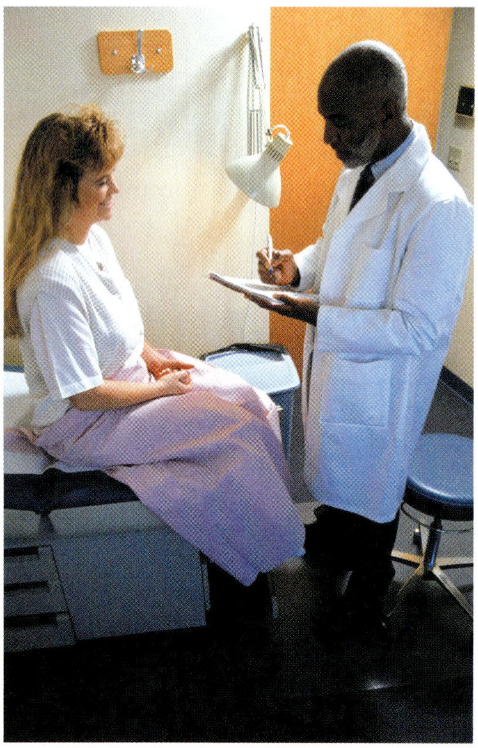

Why is it necessary to use placebos in clinical trials? For many years, scientists observed that people in clinical trials who receive placebos were more likely to show an improvement than other patients with the same disease. This improvement in the health of people receiving placebos is known as the **placebo effect**. Scientists give several explanations for the placebo effect. One factor is the regular, and often better, medical care that patients receive in a clinical trial. Another factor is the psychological effect of participating in a clinical trial. A patient who is getting regular medical attention and a pill may have a more positive attitude toward his or her health. He or she may also be more careful about diet, exercise, or other factors related to the illness. As a result, this person may be more likely to recover because of this positive attitude, and not as a result of anything in the pill. Today, scientists and doctors are studying this connection between a person's mental attitude and his or her physical health to understand how they influence each other.

STOPPING TO THINK 4

In your own words, what is the placebo effect?

The placebo effect sometimes makes it difficult to determine if a treatment is effective. For example, the use of vitamin C to treat colds is controversial. Some studies suggest that vitamin C improves cold symptoms. Other studies suggest that vitamin C doesn't work any better than a placebo, and that people who feel better from vitamin C are experiencing a placebo effect. Studies of high blood pressure, asthma, pain, depression, and cough have shown that about 30–40% of patients taking a placebo experience some relief of their symptoms. This relief can sometimes be measured objectively: for example, blood pressure actually drops in some patients taking placebos.

ANALYSIS

1. In clinical trials of medicines, why is one group of volunteers given a placebo? Explain.

2. Activity 3, "Testing Medicines: A Clinical Trial," simulated a clinical trial that investigated whether a headache medicine was effective. A person reporting a better or worse taste with the pink lemon drink (compared with the yellow lemon drink) represented a person feeling better after taking a pill for headache relief. In this simulation:

- **a.** What represented the medicine?
- **b.** What represented the placebo?
- **c.** Look at your data on Student Sheet 3.1, "Analysis of Clinical Trial." How many people experienced the placebo effect?
- **d.** Look at your data on Student Sheet 3.1. How many people in the placebo group were unaffected by the placebo?

3. Imagine a clinical trial to test a treatment for serious illnesses, such as heart disease or cancer. What is the trade-off of giving placebos to some people participating in this clinical trial?

Table 1: Clinical Trial of Cold Medicine				
	Feel the Same	**Feel Better**	**Feel Worse**	**Total Number of People in Group**
Control Group (Received placebo)	60	35	5	100
Treatment Group (Received medicine)	10	80	10	100

4. Review the data shown above from a clinical trial of a cold medicine.

a. Copy and complete a table like the one shown below to compare the number of people who feel better as a result of the medicine vs. the placebo:

Table 2: Analysis of Clinical Trial Data

	Number of people who feel better	Total number of people in group	Percent who feel better
Control group (Received placebo)			
Treatment group (Received medicine)			

b. Is the medicine effective in a high percentage of the population? Explain your answer.

c. Compare the percentage of people who feel worse in the placebo group vs. the percentage who feel worse in the treatment group. What difference do you observe? What could explain this difference?

d. Would you conclude that this medicine is safe and effective for treating colds? Explain the evidence for your conclusion.

Table 3: Clinical Trial of Headache Pill

	Feel the Same	Feel Better	Feel Worse	Total Number of People in Group
Control Group (Received placebo)	50	46	4	100
Treatment Group (Received medicine)	41	55	4	100

5. Review the data shown above from a clinical trial of a headache pill. Think about whether the medicine works and how often side effects occur. Explain why this medicine should or should not be sold. Support your answer with evidence about the safety and effectiveness of this medicine.

6. **a.** **Reflection**: What evidence do you use to decide if cold, headache, or other over-the-counter medicines work for you?

b. What effect do you think a positive attitude has on your everyday health? What about when you are sick?

5 Can You Feel the Difference?

LABORATORY

Good scientific experiments are often designed to test only a single factor, or **variable**. You may remember that Dr. Goldberger had screens put on the windows during his experiment on prisoners. He also had bedsheets and clothes washed regularly. One strength of Dr. Goldberger's experiment was his effort to reduce all the other variables—such as the presence of insects or variations in cleanliness—that could affect his results. His goal was to make diet the *only* factor that was being changed. In this case, diet was the variable being tested.

Scientists are interested in how people respond to the environment. People use their senses—touch, sight, hearing, smell, and taste—to get information about their surroundings. This information then travels through nerves to the human brain. In the next two activities you will investigate your sense of touch. Can you identify all of the variables that might affect your results? What can you do to try to keep all of these variables the same?

CHALLENGE

Which part of your arm is most sensitive to touch?

MATERIALS

For each pair of students

1 **2-point sensor**

2 **plastic toothpicks**

SAFETY

Be careful when doing the touch tests. Press gently when testing, making sure to only slightly depress the skin surface.

PROCEDURE

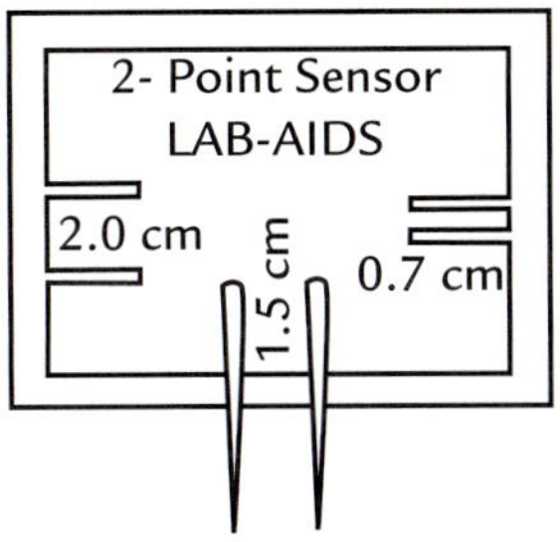

1. Slide 2 plastic toothpicks into the 2-point sensor on the side marked "1.5 cm."

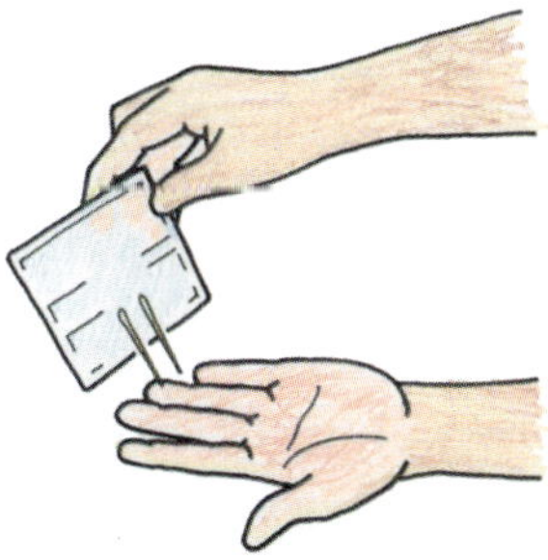

2. With your eyes open, investigate your sense of touch by touching the skin of your fingers, palm of your hand, and forearm with the point of just *one* toothpick.

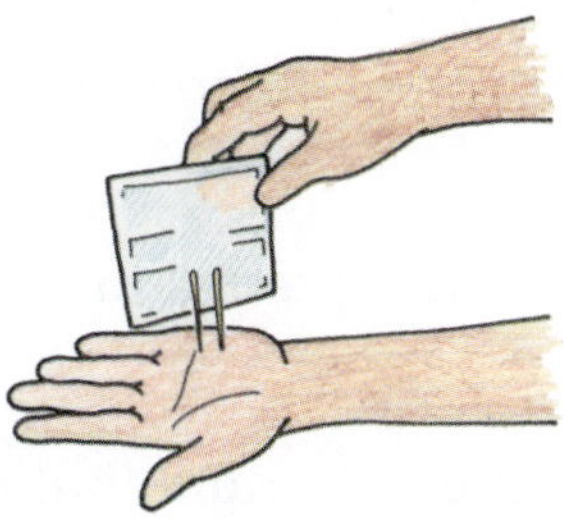

3. With your eyes open, touch your fingers, palm, and forearm with the points of both toothpicks.

4. Record your observations in your science notebook while your partner investigates his or her own sense of touch.

5. Have your partner close his or her eyes while you touch the skin on his or her fingers with either one or two toothpick points. Touch just hard enough to see that the points are barely pushing down on the skin. Randomly alternate between one and two points. Can your partner tell the difference?

6. Create a larger version of the table shown on the next page. In the table, record your observations about your partner's ability to tell the difference between one and two points on his or her fingers.

Table 1: Observations of Touch Sensitivity

Person Being Tested	Fingers	Palm	Forearm
(Name)			
(Name)			

7. Repeat Steps 5 and 6 on your partner's palm *and* forearm.

8. Switch places and repeat Steps 5–7.

9. In your group of four, use Analysis Questions 1 and 2 to discuss your results.

ANALYSIS

1. Which part of your arm—your fingers, palm, or forearm—was the most sensitive to touch? What data do you have to support your conclusion?

2. In your group, how many people found fingers to be the most sensitive part of their arm? How many found palms or forearms to be the most sensitive? How similar were different individuals' responses to touch?

3. Why was it important for the person being tested to close his or her eyes?

4. Before scientists make comparisons, it is important that they perform well-designed experiments. In a well-designed experiment, all of the variables, except the one being tested, are kept the same.

   a. In your experiment, what variables did you keep the same?

   b. Were there any variables (except for the one being tested) that you could not keep the same?

6 Finding the Nerve

You probably found that some parts of your arm were more sensitive than others. In this reading you will learn about some of the reasons for these differences.

CHALLENGE

Why do different parts of your body have different sensitivities to touch?

READING

You use your senses, to gather information about your environment. Your senses include sight, hearing, taste, smell, and touch. To understand how you feel objects you touch, you need to know a little about your nervous system. Your **nervous system** includes your brain, your spinal cord, and your nerves. Figure 1 shows how these parts of your nervous system are connected. Nerves are found throughout your body, from the tips of your toes to the top of your head.

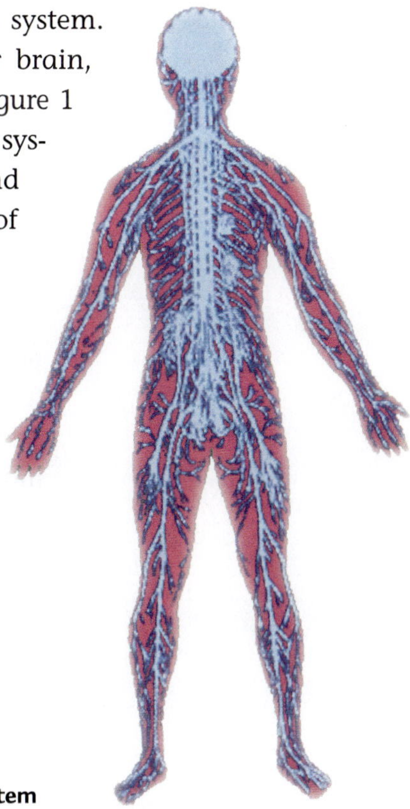

Figure 1: Human Nervous System

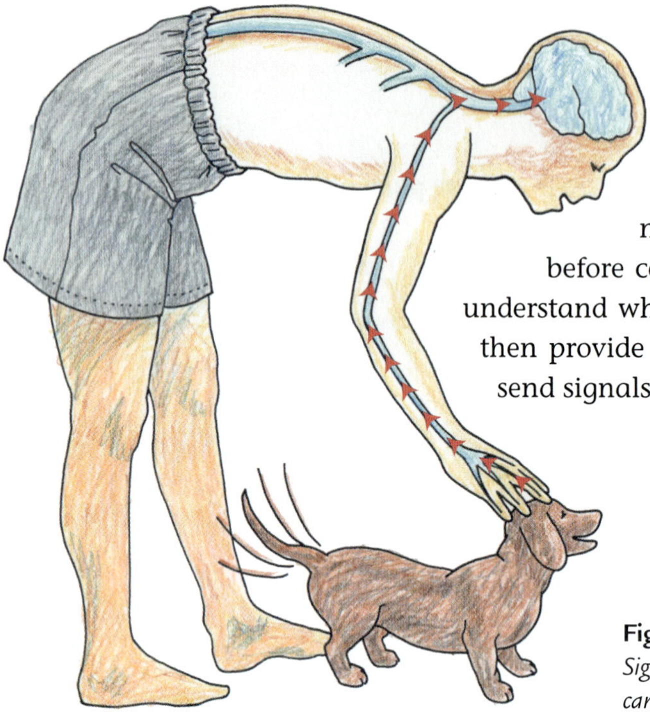

Your nerves take in information from the world. For example, your nerves have helped you collect information about the weight of your backpack when you pick it up and think, "It's heavy!" Information travels from nerves all over your body to your spinal cord before continuing to your brain. Your brain helps you understand what your nerves have detected. Your brain can then provide directions to other nerves. These other nerves send signals to your muscles that cause you to move.

Figure 2: Responding to the Environment
Signals travel from nerves in your body to your brain. These signals can travel up to several hundred miles per hour!

STOPPING TO THINK 1

a. Someone accidentally bumps into you in the hallway. What part of your body detected the feeling of being bumped?

b. Where in your body is this feeling analyzed?

Sometimes your nervous system responds to messages that you provide. For example, you decide to eat a sandwich. Your brain sends signals along nerves to direct your muscles to pick up the sandwich and take a bite. But what happens after you swallow? Do you have to think about directing your body to process the food? Your body automatically moves food through your digestive system and processes it. For this to happen, muscles inside your body must move. These muscles are controlled by the automatic, or involuntary, part of your nervous system. You don't even have to think about it!

STOPPING TO THINK 2

Is breathing completely involuntary? Explain.

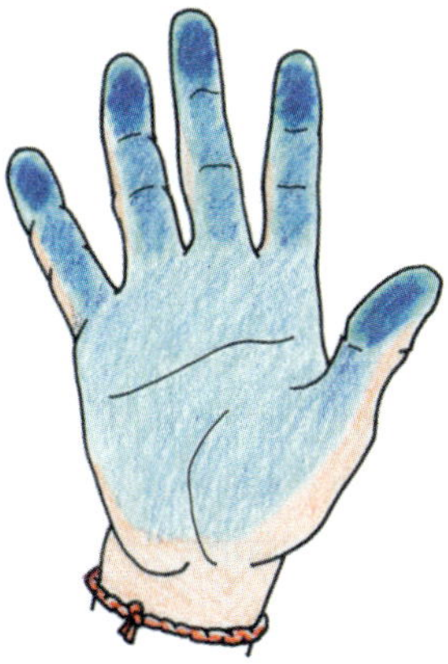

Figure 3: Touch Receptors on Human Hand

Why are some parts of your body more sensitive to two points than others? When something touches your arm, you feel it if it is detected by the nerve endings in your skin. You have nerves in your body that detect pressure, heat, sounds, smells, and light. The nerve endings that detect pressure on your skin are called **touch receptors** (ree-SEP-tors). They help carry a message from your skin to your brain.

Some parts of your body have more touch receptors than others. When two points stimulate the same touch receptors, you feel the points as one touch. When they stimulate different touch receptors, you feel two different touches. Figure 3 shows the concentration of touch receptors on a hand. Notice that the tip of the finger has more touch receptors than the rest of the finger. When you reach out to touch something, you often use your fingertips. You may have heard of people with limited or no vision reading Braille. Braille is a written language that uses raised dots instead of letters. Braille is read with fingertips. Not surprisingly, you have the greatest number of touch receptors right at your fingertips—just where they are needed.

EXTENSION

For links to more information on the human nervous system, go to the SALI page of the SEPUP website.

ANALYSIS

1. a. Where would you expect to have more touch receptors: on the palm of your hand or on the back of your hand? Explain your ideas.

b. Explain how you could test your answer to Question 1a.

2. Review your results from Activity 5, "Can You Feel the Difference?" Based on what you now know, where on your arm—fingers, palm, or forearm—do you have the fewest touch receptors?

7 Human Variation

LABORATORY

In the last two activities, you learned about your sensitivity to touch on different parts of your arm. How can you find out if what you learned about yourself applies to other people? You and your classmates will collect and compare data to find out how sensitive each of you are to touch on the palm of your hand. Do you think everyone has the same sensitivity to touch?

CHALLENGE

What is the smallest distance at which you can still feel two points? How does this compare with other people in your class?

MATERIALS

For each pair of students

1 2-point sensor
6 plastic toothpicks
1 metric ruler (optional)

For each student

1 Student Sheet 7.1, "Touch Test Data"
1 Student Sheet 7.2, "Sensitivity to Two Points: Class Results"
additional copies of Student Sheets 7.1 and 7.2 (optional)

SAFETY

Be careful when doing the touch tests. Press gently when testing, making sure to only slightly depress the skin surface.

PROCEDURE

1. *What is the smallest distance—0.7 cm, 1.5 cm, or 2.0 cm—at which you think you can still feel two points on the palm of your hand?* In your science notebook, record your hypothesis. Be sure to explain *why* you made this prediction.

2. Identify your dominant hand. (This is usually the hand you write with.) Throughout the experiment, you will test your dominant hand.

3. Begin completing Student Sheet 7.1, "Touch Test Data." Write in your name and circle which of your hands is dominant. You will begin by testing the palm of your hand, so circle "palm" as the part of the arm being tested.

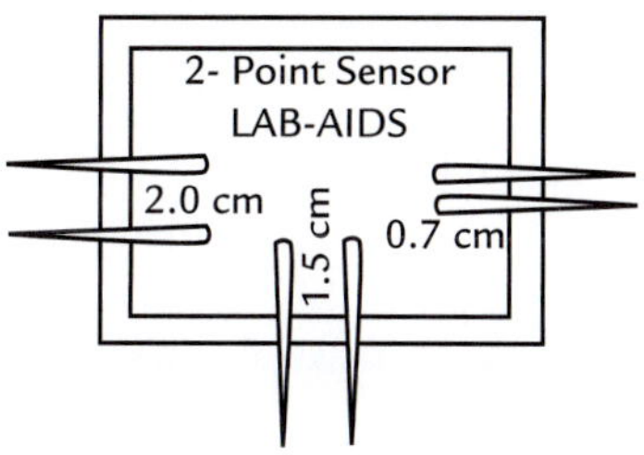

4. Since you will test your partner (and vice versa), switch student sheets so you can record his or her data on his or her sheet.

5. Slide 2 toothpicks into each side of the 2-point sensor as shown on the left. You should end up with toothpicks on three sides, with the toothpick points 0.7 cm apart, 1.5 cm apart, and 2.0 cm apart.

6. Practice using the 2-point sensor so that you can safely and easily test using any of the three sides.

7. As the experimenter, you will use the 2-point sensor to test your partner. Record your partner's responses on Table 1, "Touch Response," on Student Sheet 7.1. *It is important that you move across each row in Table 1 as you record your partner's response.*

   For example, when you do Trial 1, you will work across the first row:

   a. Turn the 2-point sensor to the 0.7 cm side and touch your partner's palm with just one point.

   b. Turn the sensor to the 1.5 cm side and touch your partner's palm with just one point.

   c. Turn the sensor to 2.0 cm and touch your partner's palm with two points.

8. Before starting the touch tests, ask your partner to close his or her eyes. *The partner being tested should not try to "guess the right answer." The goal is to report what you really feel—one point or two.*

9. Use Table 1, "Touch Response" to test your partner. *You can start with any row you want, but be sure you complete all the rows. Don't tell your partner which row you are using.* Remember to touch just hard enough to see that the points are barely pushing down on the skin. After each touch test, have your partner report whether he or she feels one or two points and record the response.

10. After you complete all the touch tests in Table 1, have your partner test you by repeating Steps 7–9.

11. Give your partner back his or her original Student Sheet.

12. Complete the rest of Student Sheet 7.1.

13. If you have time, repeat this experiment to test the sensitivity of your fingertips and your forearm.

ANALYSIS

1. **a.** According to your data, what can you conclude about your sensitivity to 2-point touches? How does this conclusion compare with your hypothesis?

   b. Compare your results with those of your partner. How similar or different are your results?

   c. Compare your results with those of another pair of students. How similar or different are your results?

2. Look at the class results on Student Sheet 7.2, "Sensitivity to Two Points: Class Results." Compare the smallest distance at which you could feel two points with the results of the rest of the class. What can you conclude about the sensitivity of different people to touch? Is it possible to make conclusions about people in general?

3. You were able to determine the smallest distance at which you could still feel two points using only the 2-point touch data. The 1-point touches acted as a control. Why would you need a control when experimenting on people?

4. **a.** A good experiment is reproducible. What parts of this experiment are reproducible?

   b. How could this experiment be improved?

5. What factors make studying people scientifically difficult? How do scientists deal with these factors in a well-designed investigation?

6. Design an experiment to determine a person's sensitivity to sound. Assume you have a machine that you can set to produce sounds of varying volume (measured in units called decibels). **Hint:** Think about how you tested your sensitivity to 2-point touches.

EXTENSION

Go to the SALI page of the SEPUP website to find out how to post your class data on the site. Look at the data posted by other students. What can you conclude about the sensitivity of different people to touch? What effect does sample size have on your conclusions? Explain your ideas.

8 Studying People

Studying people *scientifically* presents some interesting challenges. One challenge is that different people react differently to identical events. Think about the results of your experiments on touch sensitivity. You probably discovered that not everyone could feel two points at the same distance. How do you think scientists deal with this type of challenge?

CHALLENGE

How are qualitative and quantitative data used when testing a hypothesis about people?

READING

Scientists usually begin with an idea about what they want to investigate. This is also true for scientists who study people. This idea, or hypothesis, is the basis for their study. As you now know, a **hypothesis** (hi-PAH-thuh-sis) is an explanation based on observed facts or on an idea of how things work. For example, Dr. Goldberger hypothesized that there was a relationship between pellagra and diet.

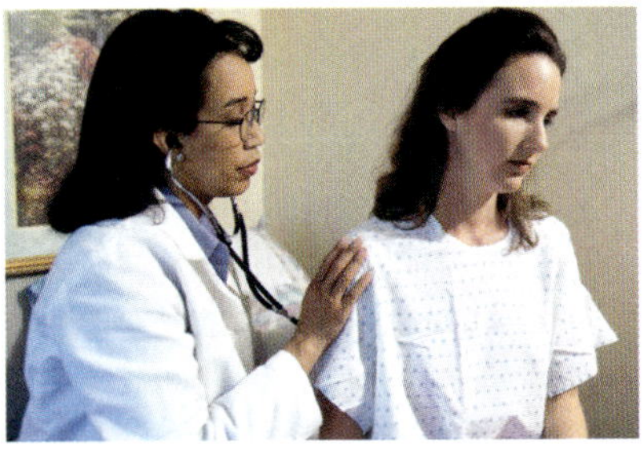

Doctor and patient

In some cases, scientists predict the possible outcomes of an experiment. A hypothesis can include a prediction of what might happen. In the last activity, your hypothesis included a prediction about the results. To make sure they don't influence results, scientists studying people often do *not* predict the possible results of their experiments.

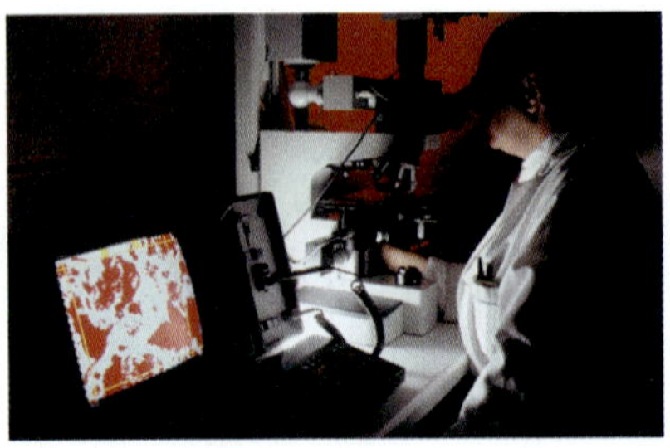

Scientist examining rat brain scan

After developing a hypothesis, scientists usually plan and conduct experiments. The results of an experiment can provide evidence for or against a hypothesis. In some cases, an experiment provides new information. This new information can produce another hypothesis and another experiment. In this way, a hypothesis can be a "work in progress" that is continually revised.

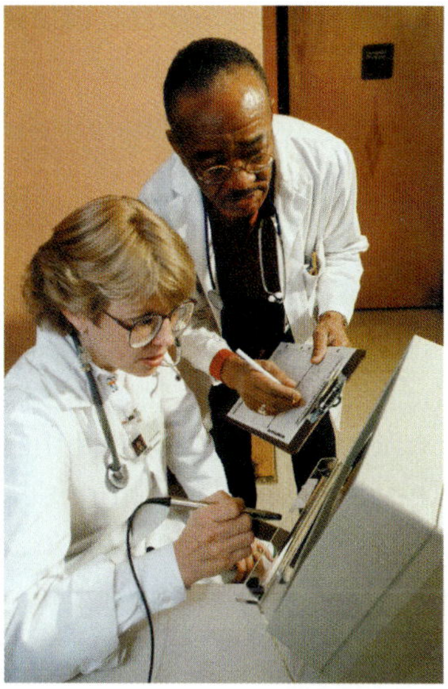

Doctor and nurse working on computer

STOPPING TO THINK 1

You hear your friend Yoshi tell someone that a hypothesis is the same as a guess. Explain whether you agree with Yoshi.

In the case of pellagra, Dr. Goldberger conducted an experiment on prisoners to test his hypothesis. During the experiment, 7 out of 11 prisoners, or 64%, developed pellagra after six months. Even though all of the prisoners ate the same diet, they did not respond in the same way. Scientists try to account for individual differences by studying a large enough **sample size**.

STOPPING TO THINK 2

Explain whether Dr. Goldberger provided enough evidence to prove that pellagra is not contagious.

*The word **data** (DAY-tuh) is plural. It refers to a set of information. The singular form is* datum, *one piece of information.*

In the United States today, government agencies, such as the Food and Drug Administration (FDA), tightly control and review experimentation on humans. After receiving government approval, companies often search for volunteers to participate in clinical trials. Clinical trials test medicines, food, medical procedures, and medical equipment. **Data** collected during these trials provide evidence for making conclusions about a product or treatment. The type of data collected depends on what is being investigated.

For example, imagine a company that is conducting trials of a new hearing aid. The company would like to know how clearly volunteers can hear when using the hearing aid. People will try out the hearing aid and report how clearly they hear sounds, and whether they hear any buzzing or other undesired noise from the hearing aid. They may also report whether they find the hearing aid comfortable to wear and attractive in appearance. These are all examples of qualitative data. **Qualitative** (KWAL-i-tay-tiv) **data** describe properties or characteristics (qualities) that are used to identify things. Qualitative data often result from putting things into categories.

STOPPING TO THINK 3

You notice a skateboard for sale on an Internet website. The price is good, but the website provides no picture or other information. What qualitative data would you like to have before you decide if you will buy it?

What if you were interested in buying the hearing aid? You would probably want to know how well it works when compared with other hearing aids. In clinical trials, people with hearing problems might be tested to measure the softest sound they could hear with and without the different hearing aids. The volume of sound (how loud it is) is measured in decibels. The company might report that in clinical tests 90% of the people wearing the hearing aid heard sounds as low as 35 decibels, when only 10% could hear these sounds without the hearing aid. In this case, numerical, or quantitative, data were collected. **Quantitative** (KWAN-ti-tay-tiv) **data** are values that have been measured or counted. The word *quantitative* is related to the word *quantity*, which means "number" or "amount."

Consider an everyday example of qualitative and quantitative information. Imagine you are at a large store with a friend when suddenly you turn around and she's not with you anymore. You look around, but you don't see her anywhere nearby. You ask a store employee if he has seen her. You describe your friend as a tall 13-year-old girl with short brown hair, brown eyes, a red jacket, two braids, and a dimple when she smiles. You have included important qualitative and quantitative information about your friend.

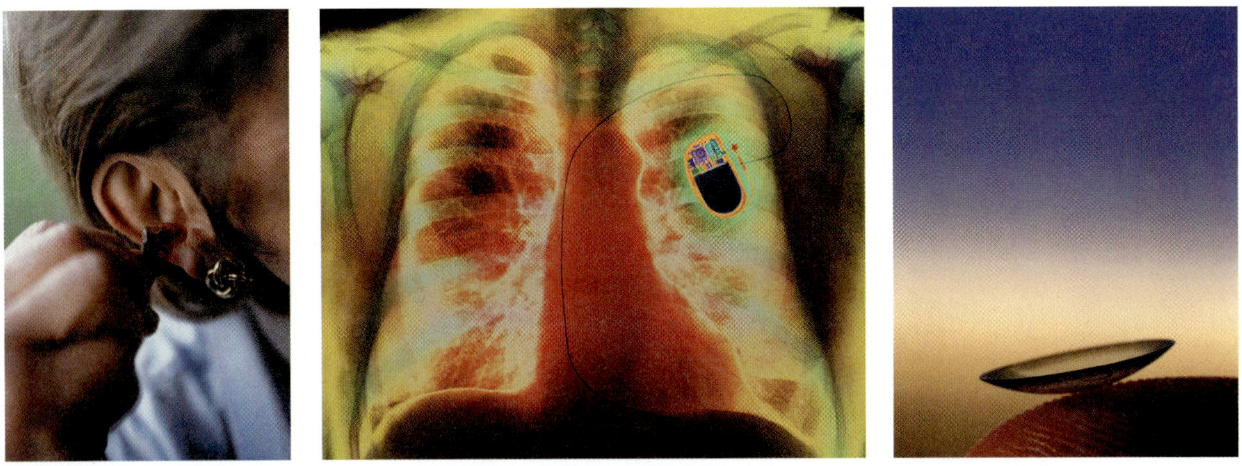

Medical products, such as hearing aids, pacemakers, and contact lenses are tested before being sold to the public.

STOPPING TO THINK 4

Reread the description of your friend in the above paragraph. Create a list of each of her characteristics. Identify each characteristic as either qualitative or quantitative.

Both qualitative and quantitative evidence are important in identifying your friend. Qualitative data provide information about important characteristics that are difficult to measure but can be described and categorized. But in some cases quantitative data help give a clearer description. Think about the description of your friend in the previous paragraph. A height that seems tall to you might seem short to the store employee. But if you tell him that your friend is 5 feet 7 inches tall, you will both have a clear idea of how tall she is.

STOPPING TO THINK 5

a. What kinds of qualitative data are useful in studying people scientifically? Provide at least two examples.

b. What kinds of quantitative data are useful in studying people scientifically? Provide at least two examples.

ANALYSIS

1. You decide to take a medicine for your upset stomach. You have a choice of two medicines. Both medicines are advertised as safe based on clinical trials. Medicine A was tested on 100 people. Medicine B was tested on 10,000 people.

   a. Which medicine would you take? Explain. Support your answer with evidence.

   b. Was your decision based on qualitative or quantitative information?

2. You're a volunteer at a local hospital. While you are there, you read a patient chart containing the data below. Identify each item of patient data as quantitative or qualitative.

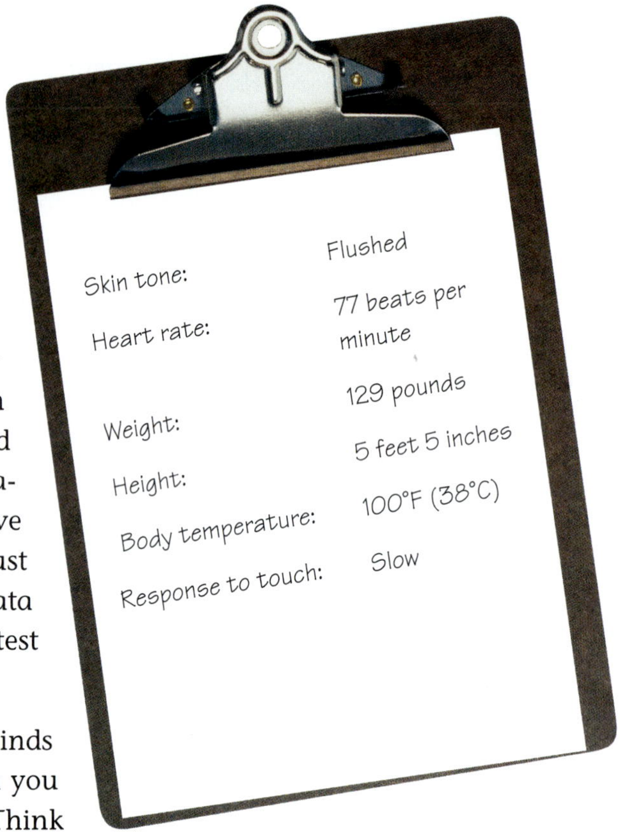

3. Imagine conducting a clinical trial of a headache medicine. Based on your research, you hypothesize that the medicine will successfully treat headaches in people. Before the Food and Drug Administration (FDA) will approve your study, you must explain the type of data you will collect to test your hypothesis.

   a. List at least three kinds of qualitative data you will collect. **Hint:** Think about what information you would collect from the volunteers and what information you would collect about the medicine in order to determine the safety and effectiveness of the medicine.

   b. List at least three kinds of quantitative data you will collect. **Hint:** Think about what information you would collect from the volunteers and what information you would collect about the medicine in order to determine the safety and effectiveness of the medicine.

4. Think about the activities you have done so far in this unit. What are the common elements of a well-designed experiment?

5. **Reflection:** Both qualitative and quantitative data provide evidence for making decisions. How have you used each of these types of data to make decisions? Describe your experiences.

9 Data Toss

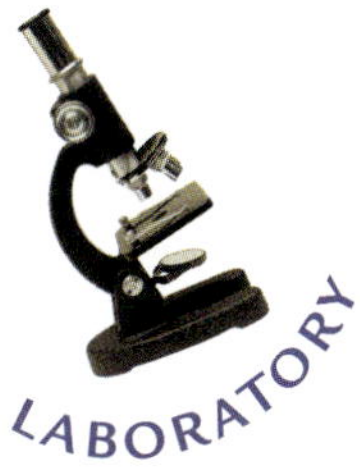

Because of differences among individuals, a **range**, or set of values, is often used to generalize to a larger group of people. For example, you may have had nurses take your pulse. What can your pulse tell them? Through scientific study, a normal pulse range for people at rest has been established. The normal range for adults is 60–80 beats per minute. The use of a range (60–80) instead of a specific number helps take into account the differences among people.

To establish a range, data are collected from many trials. Increasing the number of people tested increases the chance that the conclusions are true for most people.

If you play baseball, softball, or other games that involve throwing and catching, your teacher or coach may tell you to catch the ball with two hands. Does catching with two hands (as compared to one) really increase your ability to catch a ball?

CHALLENGE

How can you study people scientifically? Find out by collecting data on people and then designing your own experiment.

MATERIALS

For each pair of students

1 **porcupine ball**

For each student

1 **Student Sheet 9.1, "Data Tables for Part One"**

PROCEDURE

Part One: One vs. Two Hands

1. Based on what you know about catching balls, record a hypothesis about whether the ability to catch a ball increases when using two hands (vs. one).
2. With your partner, brainstorm all of the variables that you will try to keep the same while conducting the experiment.
3. Stand 2–3 meters away from your partner, as directed by your teacher.
4. Have your partner toss you the ball while you catch it using only one hand. Remember to keep all variables, except for the one being tested, the same.
5. *Did you find catching the ball with one hand easy, difficult, or somewhere in between?* Record your response in Table 1, "My Data," of Student Sheet 9.1, "Data Tables for Part One."
6. Have your partner toss you the ball 20 times while you continue to catch it using only *one hand*. Record the results of each catch in Table 2, "My Data: Number of Completed Catches." Place an "X" in the box when you catch the ball using one hand, and an "O" when you do not. When you are done, count and record the total number of completed catches.
7. Repeat Steps 3–6, but use *two hands* to catch the ball.
8. Now repeat Steps 3–7 for your partner to collect his or her own data.

9. Share your results with the class. Use the class data to complete Tables 3 and 4 on Student Sheet 9.1.

10. Complete Analysis Questions 1–4.

Part Two: Designing Your Own Experiment

11. Design an experiment to investigate the effect of one variable on your ability to catch a ball. For example, you may want to investigate your ability to catch with one eye closed, to compare your ability to catch with your right or your left hand, or to vary the size of the ball. When designing your experiment, think about the following questions:
    - What is the purpose of your experiment?
    - What variable are you testing?
    - What is your hypothesis?
    - What variables will you keep the same?
    - How many trials will you conduct?
    - Will you collect qualitative and/or quantitative data? How will these data help you to reach a conclusion?
    - How will you record these data?

12. Record your hypothesis and your planned experimental procedure in your science notebook.

13. Make a data table that has space for all the data you need to record. You will fill it in during your experiment.

14. Obtain your teacher's approval of your experiment.

15. Conduct your experiment and record your results.

16. If you have time, switch procedures with another pair of students. Conduct their experiment and record the results while they do the same with yours. Then exchange results so that each pair has two sets of results for the experiment they designed.

EXTENSION

Conduct your experiment with different age groups, such as children vs. adults. Compare your data with the data collected in class. Are your conclusions the same?

ANALYSIS

Part One: One vs. Two Hands

1. Explain whether your own data supported or disproved your hypothesis about the ability to catch a ball with two hands (as compared to one).

2. Look at the class data on Student Sheet 9.1.

    a. In Table 4, what was the range of students' ability to catch a ball with one hand?

    b. In Table 4, what was the range of students' ability to catch a ball with two hands?

    c. Use the class data to explain whether the ability to catch a ball increases when using two hands (as compared to one).

3. **a.** What qualitative data did you collect in Part One of this activity?

    b. What quantitative data did you collect in Part One of this activity?

    c. Which type of data was more useful for comparing results among the class members?

4. Your coach asks you whether she should continue to recommend that players on the team catch with two hands. Explain your recommendation and whether it is based on your own data or that of the class.

Part Two: Designing Your Own Experiment

5. What conclusions can you make based on the results of your experiment? Explain how your conclusions are based on the data collected during your experiment, and whether your hypothesis was supported or disproved.

6. **a.** In your experiment, what variables did you keep the same?

   b. Were there any variables (except for the one being tested) that you could not keep the same?

   c. How could you improve the design of your experiment? Explain.

7. Was it possible for another pair to repeat your experiment? Were the results similar to your original results? Explain.

8. How do scientists use good experimental design to collect reproducible data about people?

10 Evaluating Clinical Trials

Conducting studies on people is time-consuming and expensive. Researchers need funding. The U.S. National Institutes of Health (NIH) is one organization that provides money for such studies. What does NIH look for when researchers request money to study people scientifically?

THE NIH COMMITTEE

Listed on the following pages are proposals for five studies requesting money to conduct clinical trials. Most of the studies are based on actual research projects. As a member of the NIH review committee, your job is to evaluate the experimental design of each proposed study. Based on your review, make a recommendation as to whether NIH should fund each study.

CHALLENGE

Which proposals have an experimental design worth funding?

MATERIALS

For each student

1 **Student Sheet 10.1, "Reviewer Comments"**

PROCEDURE

1. With your group, brainstorm all of the factors that are important in designing an experiment to study people scientifically. Work together to agree on a complete list.
   - Listen to and consider the explanations and ideas of other members of your group.
   - If you disagree about a factor with others in your group, explain to the rest of the group why you disagree.
2. Read one of the studies. (The studies begin on the next page).
3. Use your list of experimental design factors from Step 1 and Student Sheet 10.1, "Reviewer Comments," to evaluate the study. Record the name of the study, the factors of its experimental design, and any other comments you have.
4. Based on your comments, decide whether you would recommend funding for this study.
5. Repeat Steps 2–4 for the remaining studies.

Study 1:
Clinical Trial of "Summer Fever" Medicine

We are working to prevent "summer fever," a contagious disease that kills or paralyzes many children each summer. This disease is spreading quickly across the United States. We have developed a medicine that we believe will prevent children from catching this disease. We would like to conduct a clinical trial to find out if our medicine is effective in children. Our plan is to give the medicine to 12,000 children in the state's capital city. We will count how many of these children catch summer fever. We are sure that all of these children will be completely protected as a result of taking this medicine.

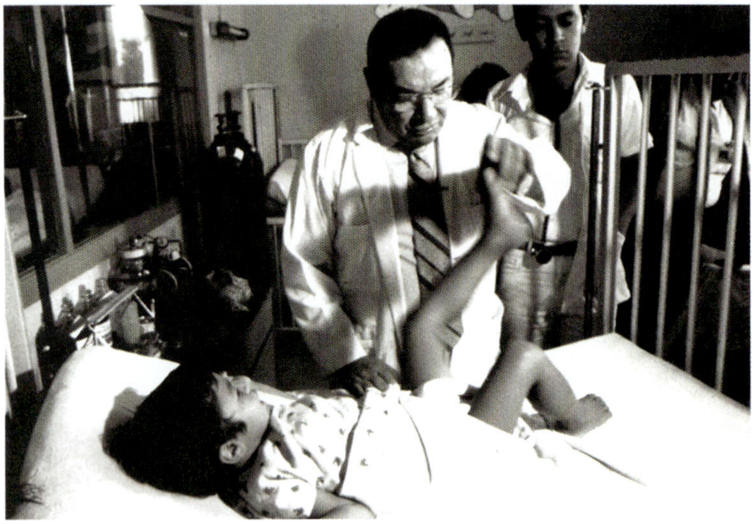

This child's legs are paralyzed as a result of a disease similar to "summer fever."

To prove that the medicine is safe for humans, my assistant, my 14-year-old child, and I have all taken it. We have not suffered any bad effects from the medicine.

Study 2:
Clinical Trial of Burn Cream

We have developed a new, medicated cream that helps heal burns on animals. Faster healing reduces pain and reduces the chance that the burns will become infected. We would like to test this cream on humans to see if it helps heal burns faster. We plan to conduct clinical trials on 200 people ages 20–40. Half of the people will be men and half will be women. Each person must have third-degree burns that are no larger than 3 centimeters by 3 centimeters. They must also be healthy except for their burns.

Fifty of the men and fifty of the women will be treated with our new medicated cream. The other half will receive a placebo cream that does not contain the medication. We will not let either group know which cream they are receiving. Each person will apply 2 grams of cream to the burn every morning for 20 days.

Three studies done by the local veterinary school have shown that the medicated cream is effective in healing burns on rabbits, cats, and dogs. Each study involved 100 animals. Half of the animals received medicated cream and half received cream without medication. The burns on animals treated with medicated cream healed faster.

Study 3: Clinical Trial of Weight Loss Method

More than half of the adults in the United States are overweight. Many of them try risky fad diets to lose weight. One way to help people lose weight is to find effective ways in which they can reduce the number of calories they eat. For this reason, we would like to conduct a clinical trial of one method of reducing calorie intake. We would like to find out whether drinking water with an appetizer before a meal will help people reduce the number of calories they eat.

We will begin with 24 volunteers who will be told about all aspects of the trial. We will select volunteers who are men between the ages of 35–55 and who are at least 20% overweight. In our control, each volunteer will eat a chicken-rice appetizer that contains exactly 200 calories. We will then count how many calories the volunteer eats for lunch after eating this appetizer. To measure the effect of drinking water with the appetizer, the same volunteer will eat the same appetizer on a different day, but this time with a 12-ounce glass of water. We will again count the number of calories the volunteer eats for lunch afterward.

Each volunteer will repeat the control and the experiment three times. We will then average the results to determine any change in the number of calories eaten in each part of the trial.

Study 4: Clinical Trial of Relaxin

We have developed a new drug called Relaxin. We believe Relaxin can be used to calm and relax people as well as prevent nausea. For these reasons, we think this drug would be useful to pregnant women. We would like to conduct a clinical trial to find out if this drug is effective for pregnant women. We will provide Relaxin to some doctors. These doctors will be asked to prescribe Relaxin to

pregnant women who ask for a drug to prevent nausea. Of course, we will tell the women that they are getting Relaxin, because people testing new drugs should give their permission. We will then record the effects on these women by asking them if they feel calmer and less nauseated.

Relaxin has been tested for its effectiveness and safety. Our scientists conducted an experiment on mice. Half of these mice were given Relaxin, while the other half were given a placebo. We observed and recorded how much the mice moved. The mice that were given Relaxin moved less than the mice that were given the placebo. This showed that the drug is effective. None of the mice had any side effects from this drug. To further test the safety of Relaxin, we gave large doses to some mice. None of these mice showed any permanent side effects from the large doses.

Study 5: Clinical Trial of Insulin

Today, insulin is used to treat diabetes. How did scientists discover how to control the effects of this disease? Watch a segment from the video, *A Science Odyssey*: "Matters of Life and Death," to find out. Evaluate the quality of the experiments shown in this segment. What other comments do you have about these experiments?

ANALYSIS

1. Which study—1 through 4—had the best experimental design? Explain.

2. Discuss your funding recommendations with your group. Do you agree on which studies, if any, should be funded? What other concerns do you have?

3. You find out that NIH has only enough money to fund one study and plans to fund the best one. Explain which study you would fund. Support your answer with evidence and identify the trade-offs of your decision.

   Hint: To write a complete answer, first state your opinion. Provide two or more pieces of evidence that support your opinion. Then consider all sides of the issue and identify the trade-offs of your decision.

4. Choose one of the studies. Review your comments about its experimental design and think about how the study could be improved. Rewrite the study to include your recommendations for improving the quality of the experiment.

5. **Reflection:** Based on what you have learned in this unit, how do scientists solve problems?

EXTENSION

Bring in news articles that describe scientific studies involving people. Analyze the studies according to the procedures you used in this activity. What issues do scientists face when studying people scientifically?

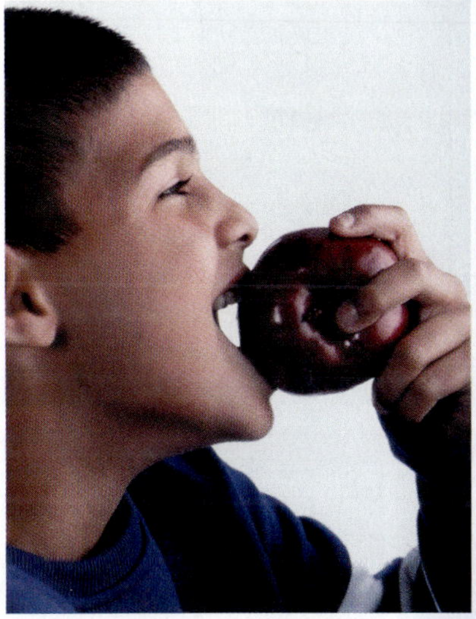

Body Works

B

Unit B

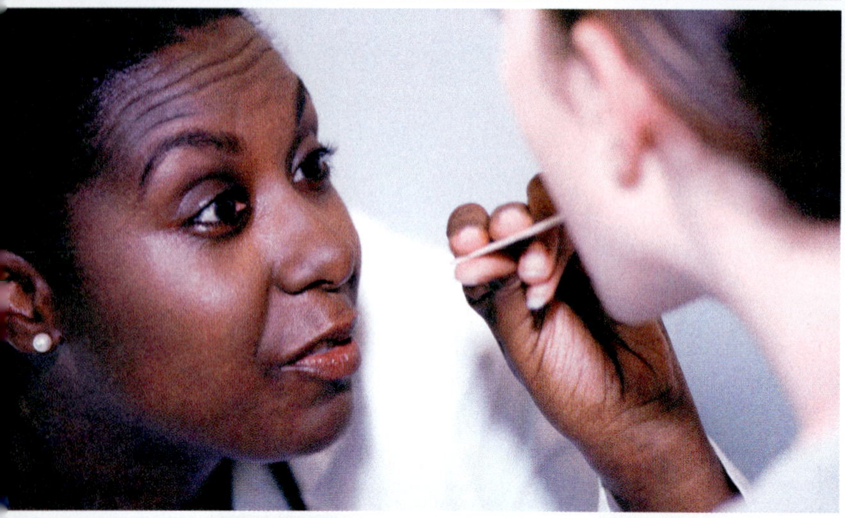

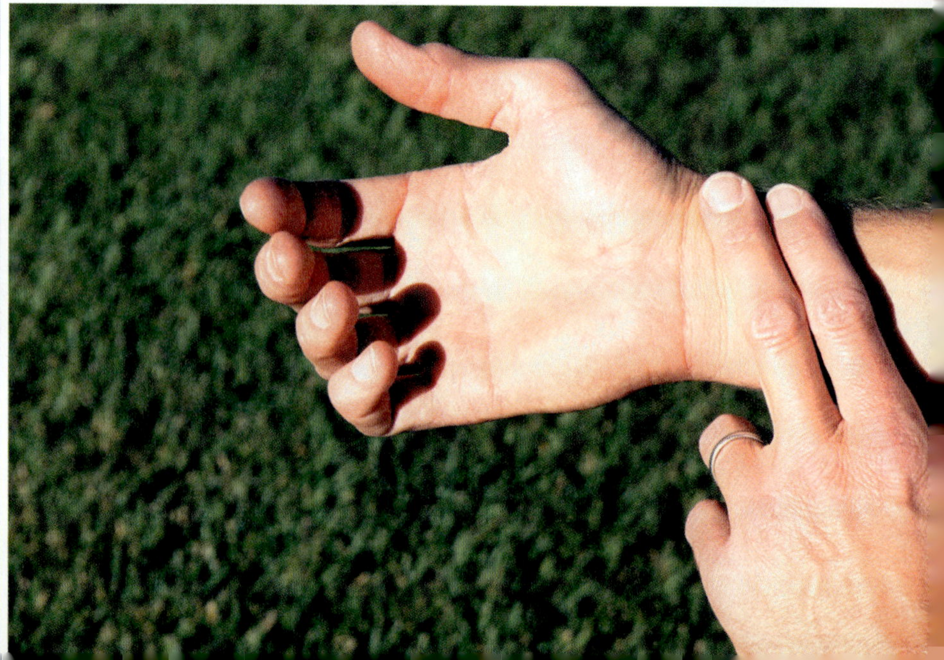

Body Works

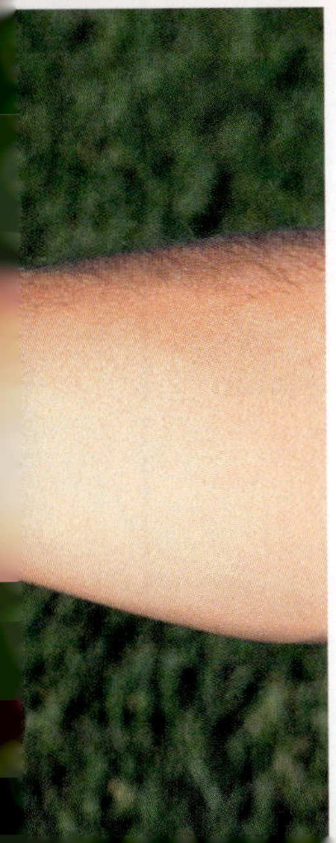

Hey Dad, did you fill out all those forms for the race?" asked Maya.

"Yes, Maya," her father replied. "The race organizers wanted to make sure that we were both in good health, so I had to answer questions about our health and list medications that either you or I were taking."

"Why do they need so much information? It's only a 5K race," said Maya.

"They need to make sure that they can take care of all the people who run in the race," explained her father. "Even though it's been several years since my heart surgery, I still take medications for my heart. I have to be a lot more careful with my health than I was before. I'm glad that they'll have all of this information, just in case."

"I sure wish you had never had any heart problems," sighed Maya.

"Me too," replied her father. "I could have taken better care of my health when I was younger, but I didn't. While it might not have prevented my heart problems, it would have reduced the risk. I hope that, as you get older, you'll make better choices than I did."

• • •

What choices do you make about your health every day? Which of these decisions may affect your health in the future? What types of information could help you make better decisions about your body? Would you be willing to fund scientific research to answer these kinds of questions?

Answering such questions thoughtfully requires knowing and understanding scientific information about the human body. In this unit, you will learn more about your own body and how it works.

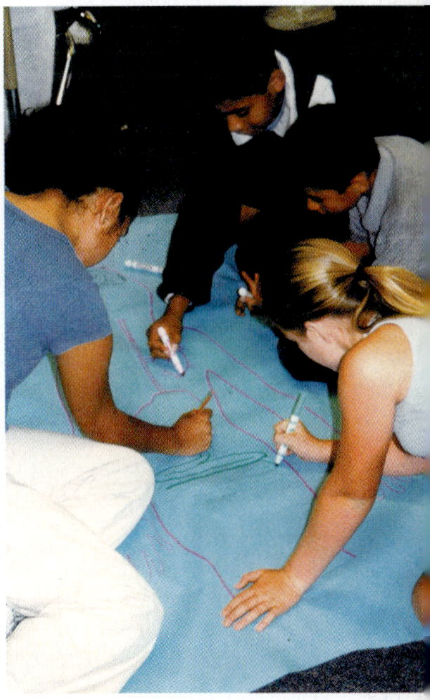

11 Sick Day

You may feel healthy most of the time. But when you feel sick, you may turn to medicines for relief. How do you decide whether to take a medicine, and what kind of medicine to take?

FEELING SICK

Sam and his older brother and sisters are talking one Saturday morning.

Sam: I feel so sick. It's just not fair! Yesterday I felt fine, but today I feel awful. I don't want to miss the birthday party at the twins' house tonight.

Andrea: You must feel pretty bad to talk about missing a party. What's wrong?

Sam: You don't really want to know. My stomach hurts so much it feels like an elephant sat on it. I had to get up and run to the bathroom four times last night and now I feel dizzy and weak.

Marta: Sounds pretty bad. Let me feel your forehead....Oh yeah, you are hot. You probably have a fever.

Andrea: Here's the thermometer. Let's take your temperature.

Robert: Don't we have something Sam can take to feel better?

Marta:	Hmm, you have a fever of 101.2°F (38.4°C). Let me check the medicine cabinet.
Andrea:	I say aspirin is the best answer. It always works for me when I get a headache.
Sam:	It doesn't always work for me when I have a headache.
Robert:	Uncle Richard takes an herbal remedy. I forget what he calls it. He says it works for him and it's safer than medicines.
Andrea:	Yeah, but Aunt Jan says herbal remedies aren't tested the way medicines are. She says you can't be sure they work and if they're safe.
Marta:	We have fever medicines, herbal tea, and some stomach medicine. I could also give you some ginger ale or soup. That's what Mom always has when she is sick.
Robert:	Why don't we get them and look at the labels?
Marta:	Good idea. And I'll call the doctor and describe your symptoms. Dad left the number by the phone.

CHALLENGE

What are the trade-offs of taking a medicine when you feel sick?

MATERIALS

For each student

1 **Student Sheet 11.1, "Notes on Medicines"**

PROCEDURE

1. Imagine you have the same symptoms Sam has. In your science notebook, record what you would do, or have done, in Sam's position.
2. Have each member of your group share what he or she wrote. Listen to each person, but do not comment.
3. Read the doctor's response to Marta to learn more about Sam's choices:

   "It sounds like Sam has a common stomach virus. He should feel better in a day or two as long as he drinks plenty of liquids. It's important that he doesn't get dehydrated. If his fever doesn't get any worse, Sam can decide for himself whether he would like to take any medicine.

   "Fever and diarrhea are sometimes ways that the body fights off bad germs. They're the body's way of trying to make you better. But a fever reducer or diarrhea medicine may help Sam feel better now. It's up to him. I'm leaving town for a couple of days, so call the hospital if his fever goes up or if he doesn't feel better in two days."
4. Shown below are the labels from the medicines that Robert found in the medicine cabinet. Use Student Sheet 11.1, "Notes on Medicines," to help you analyze the labels. Read them carefully as you decide what Sam should do.

Medicine A: Pain Reliever/Fever Reducer Tablet

Warning: If you consume three or more alcoholic drinks every day, ask your doctor whether you should take this product or other pain relievers/fever reducers. This product may cause liver damage. **Do not use** with other pain killers/fever reducers unless directed by a doctor. Keep this and all drugs out of the reach of children.

Stop using and consult a doctor if:
· symptoms do not improve · new symptoms occur · pain or fever persists or gets worse · redness or swelling is present

Medicine B: Pain Reliever/Fever Reducer Tablet

Warning: Children and teenagers should not use this medicine for chicken pox or flu symptoms before a doctor is consulted about Reye's Syndrome, a rare but serious illness associated with this medicine. Do not take this product if you are allergic to aspirin or if you have asthma, unless directed by a doctor.

Drug Interaction Precaution: Do not take this product if you are taking a prescription drug for anticoagulation (thinning the blood), diabetes, gout, or arthritis, unless directed by a doctor.

Medicine C: Diarrhea Tablet

Warning: Keep this and all drugs out of the reach of children. Do not use for more than two days unless directed by a physician. Do not use if diarrhea is accompanied by high fever (greater than 101°F), or if blood or mucus is present in the stool, or if you have had a rash or other allergic reaction to this medication. If you are taking antibiotics or have a history of liver disease, consult a physician before using this product.

Medicine D: Herbal Tea

This plant grows wild in the United States, and Native Americans have used it for centuries for a variety of purposes. This tea contains a blend of leaves, flowers, and stems of an organically grown plant. In modern studies, leaves, flowers, and roots of this plant have shown measured effects in supporting the immune system.*

*The Food and Drug Administration has not evaluated these statements. This product is not intended to diagnose, treat, cure, or prevent any disease.

ANALYSIS

1. What kinds of information are provided on the labels?

2. Sam decided to make a decision based on the information he has. Think about all of the options available to Sam. He could

   a. drink liquids and wait until he feels better,

   b. drink liquids and take one of the medicines, or

   c. drink liquids and take a combination of the medicines.

   If you were Sam, what would you do? Assume that Sam's medical history is the same as your own. Support your answer with evidence and identify the trade-offs of your decision.

   Hint: To write a complete answer, first state your opinion. Provide two or more pieces of evidence that support your opinion. Then consider all sides of the issue and identify the trade-offs of your decision.

3. Explain whether you used more qualitative or quantitative evidence to make your decision. Support your answer with examples.

4. What are the trade-offs of taking a medicine when you feel sick?

12 What's Happening Inside?

INVESTIGATION

Organs in the human body are divided into **systems** based on their function. For example, consider the excretory (ECK-skruh-tor-ee) system. The function of the excretory system is to remove liquid waste from the body. The kidneys help perform this function, so they are a part of the excretory system. You may be familiar with the organs and functions of other systems, such as the digestive or circulatory systems. Use your knowledge of the human body to look more closely inside yourself. In the photos below, which organ systems help each of the students do the activities shown?

CHALLENGE

What do you know about the human body and how it functions?

MATERIALS

Part One: Laying It Out

For each group of four students

1 sheet of chart paper
4 markers of assorted colors
glue or tape (optional)

For each student

1 Student Sheet 12.1, "Functions of Human Body Systems"
scissors (optional)

Part Two: Modeling the Human Body

For each group of four students

4 differently colored sticks of modeling clay
1 human body torso model

For each student

Student Sheets 12.2a and 12.2b, "Human Body Systems"

PROCEDURE

Part One: Laying It Out

1. With your group, draw an outline of a human body on a piece of chart paper.
2. Have each person in your group take a marker. Work together to do the following:
    - **a.** Draw in the major internal organs of the body.
    - **b.** Write down what you think each organ does.
    - **c.** Around your outline write questions you have about the human body.

3. Go around the room to view the drawings of other groups.

4. Use your own knowledge as well as information from the Introduction and Table 1 to work by yourself to complete Student Sheet 12.1, "Functions of Human Body Systems."

Part Two: Modeling the Human Body

Table 1: Organs and Structures to Model

Clay Color	Organs and Structures
	muscles of back and buttocks
	spinal cord
	kidneys (connected to bladder by thin tubes)
	esophagus
	stomach
	small intestine
	large intestine (and rectum)
	liver
	windpipe (trachea)
	lungs
	heart
	bladder
	rib cage (ribs and sternum)

5. Place the human body torso model on a flat surface.

6. As a group, use Student Sheets 12.2a and 12.2b, "Human Body Systems," and the modeling clay to make a 3-dimensional **model** of some of the organs and structures in the human body. In this model the plastic represents skin.

7. Use the modeling clay to create each of the organs listed in Table 1. Student Sheets 12.2a and 12.2b can help guide you in forming the organs correctly.

8. Place the organs into the bottom half of the human body torso model. Follow the order listed in Table 1 by placing the first structure (the muscles) down first and then adding the others in the order listed. Remember to use Student Sheets 12.2a and 12.2b for help. **Hint:** You'll need to put the end of the digestive system behind the bladder to make it accurate.

9. When you are done modeling the organs, place the other half of the plastic model on top of the body. You have now created a model of your internal organs.

10. Compare the placement of the internal organs in the model to your own body. Try to figure out where these organs are in your body.

11. Take your model apart. Roll the modeling clay back into separate balls of each color.

➢

ANALYSIS

1. The liver is the largest internal organ of the human body. Was the liver the largest organ in your model? Do you think that the other organs you modeled were accurate in size? Why or why not?

2. Was the model that you created in Part Two a good model of the human body? Why or why not?

3. Prepare a table with headings as shown below. Fill in the first column with the organs or structures listed in Table 1.

   a. In the second column of your table, identify the system that matches each organ or structure. For example, the stomach is a part of the digestive system.

   b. In the third column of your table, identify the function of each of the systems you mentioned in 3a.

Organs and Structures	System	Function

4. **Reflection:** What new things have you learned about the human body in this activity?

13 Living With Your Liver

How often do you think about what's going on inside your body? Most healthy people don't need to worry about what's happening inside. But knowing more about the human body can help you make better decisions about your health.

• • •

For links to more information about the liver, go to the SALI page of the SEPUP website.

CHALLENGE

How does the liver help your body stay in balance?

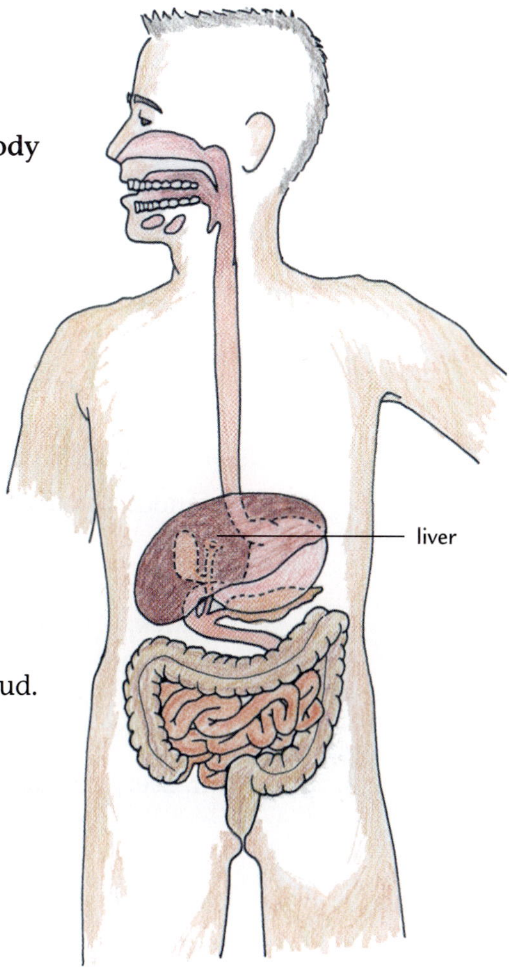

PROCEDURE

1. Assign a role for each person in your group. There are four roles: Mr. Lee, Rick, Kamika, and Yolanda. Assuming there are four people in your group, each of you will read one role.
2. Read the following role play aloud.

LIVING WITH YOUR LIVER

Rick: Mr. Lee, you told me that we're going to study the digestive system soon. I think it's going to be pretty interesting.

Yolanda: I think digestion is gross. My mom's a surgeon and she showed me some pictures of the inside of...

Kamika: Relax, Yolanda. Rick cares about the digestive system because he almost died from something he ate this summer.

Mr. Lee: Almost died? What happened, Rick?

Rick: My little sister and I found some mushrooms growing near our house. I ate some. She doesn't like it when Dad puts mushrooms in our salads, so she didn't eat any.

Yolanda: My grandpa says you should never eat wild stuff whose name you don't know.

Rick: Well, I sure won't do it again. I never felt so awful. The nurse in the emergency room said I was lucky my dad brought me in so fast.

Kamika: So the mushrooms burned your stomach?

Mr. Lee: Actually, most poisons are dangerous because they can destroy your liver.

Yolanda: Why is that? I thought the liver was just one of those weird organs that doesn't really do much.

Rick: Nope. It does a whole lot of stuff your body can't live without.

Yolanda: Isn't the liver huge? Isn't it the size of your brain?

Mr. Lee: That's close, Yolanda. In fact, it's even bigger! A liver weighs over 3 pounds. It's the largest organ in your body, except for your skin.

Kamika: So Rick almost died because those mushrooms hurt his liver? That doesn't make sense. The food you swallow doesn't even go to the liver.

Mr. Lee: That's true, but after the food is broken down, your blood carries the substances you've digested to the liver first. The liver controls which substances get stored or filtered out. Only then are these substances carried to the rest of your body.

Rick: Oh, I get it. The liver's sort of like a traffic cop that regulates which cars go and which cars stop.

Mr. Lee: Exactly. That's a great metaphor, Rick!

Yolanda: I remember my mom's friend talking about this—she's a toxicologist. The liver has to break down harmful substances so they don't get to the rest of your body and hurt other organs.

Kamika: Now I get it. Rick's liver had so much toxic stuff sent to it all at once that it got damaged.

Mr. Lee: You're right, Kamika. Even now Rick probably has to be careful what he eats while his liver recovers.

Rick: Yeah. That's because the liver also helps you digest fats, and helps regulate how much cholesterol (kuh-LESS-tuh-rall) and sugar are in your blood.

Yolanda: It sounds like the liver controls how much and what kinds of substances go to different organs and systems.

Mr. Lee: Yes, the liver helps your body keep in balance. That's what **regulation** is—keeping things balanced and responding to changing needs. The liver does many things in your body, but most of them involve regulation.

Kamika: What's a healthy balance of a mushroom poison in your blood? Zero, I bet!

Rick: I think so!

Kamika: But maybe that's not true for cholesterol and sugar. Wouldn't you need some around all the time so your body can use it?

Yolanda: Yeah, my mom says you need some cholesterol. *Too much* cholesterol is the problem.

Mr. Lee: And sugar is what we use as a quick source of energy, but too much or too little in the blood can be a serious problem!

Rick: Mr. Lee, I overheard a doctor telling my parents that if I were an adult, he would tell me not to drink any alcohol while my liver was recovering.

Yolanda: Why would that be?

Kamika: Maybe it's harmful!

Mr. Lee: That's right, Kamika. It is harmful. We call these harmful substances toxins. A **toxin** (TOX-in) is any substance that can cause damage to your body. Alcohol is a toxin. It can cause a lot of damage if someone drinks a large amount all at once or smaller amounts over long periods of time. If someone's liver is already damaged, alcohol can be toxic in even smaller amounts.

Rick: So your liver can wear out, but a little bit at a time.

Mr. Lee: That sounds right, Rick. Sometimes damage to the liver builds up over many years as the liver works and works to remove toxins. Scar tissue forms, which is called cirrhosis (si-ROW-sis). If it's bad enough, you need a transplant.

Kamika: Cirrhosis! That's what my cousin has! I didn't realize it meant a worn-out liver. But she's only 19 and she doesn't touch alcohol. She has hepatitis (hep-uh-TIE-tus). She got it from a blood transfusion when she was a baby.

Mr. Lee: Your cousin must have hepatitis C. Until about 1990 there was no way to check donated blood to find out if it was infected. There's still no vaccine for hepatitis C, but there is for hepatitis A and hepatitis B.

Yolanda: Hepatitis attacks the liver. They made sure to vaccinate my uncle for hepatitis A and B as soon as they saw he had liver damage.

Rick: So Kamika's cousin won't ever be allowed to drink, I guess. Hey, you know what other things are toxins? Ibuprofen (eye-byoo-PRO-fin) and acetaminophen (uh-see-tuh-MIN-uh-fin), those headache medicines. I had to stay away from them when I was sick.

Yolanda: Wow, I don't think of medicines as toxins.

Kamika: They do have side effects. I heard that if you take too large a dose of just about anything it can reach toxic levels.

Mr. Lee: In many cases, the effects of medicines—good and bad—would last a lot longer if the liver didn't work so hard at breaking them down quickly.

Rick: I'm just glad I'm gonna make it without a liver transplant.

Kamika: It's a good thing, because there's a shortage of organs. I know because they put my cousin on a waiting list just in case. I think kidneys are a little easier to get.

Mr. Lee: Half of the 20,000 or so transplants done in the United States each year are kidney transplants. About one quarter are liver transplants.

Yolanda: Kidney transplants must be more common because a living person can donate a kidney, since you need only one to survive.

Rick: Well, we could never survive with just half a liver!

Yolanda: Yeah, but the liver can do this cool thing. If you take a dead person's liver and put half of it into two people who need livers, the two halves, uh...

Mr. Lee: Regenerate (rih-JEH-nuh-rate). The halves grow back into complete livers. No other complex organ can regenerate. For example, the heart and the brain can't do it.

Rick: That means a living donor can give half a liver, and the half still left will regenerate.

Kamika: That's really amazing! I can't wait to tell my cousin about regeneration, in case she needs a transplant someday.

ANALYSIS

1. What are some of the functions of the liver?

2. People who have cirrhosis of the liver are usually on a strict diet. They have to be careful of what they eat and drink. Why do you think this is?

3. How can understanding how your body works help you make decisions about your health?

14 Breakdown

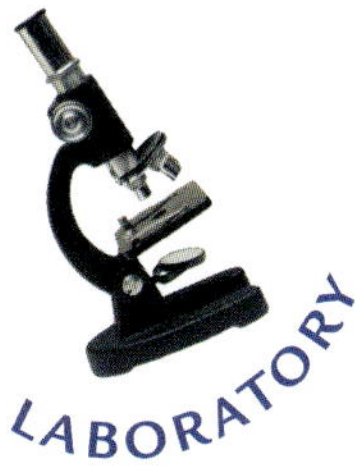

You already know the organs in the digestive system. But what exactly do they do? One important function of the digestive system is to break down food into smaller pieces. Only then can the nutrients in the food be absorbed by your body.

When you chew food, **mechanical breakdown** occurs. Most mechanical breakdown occurs in your mouth with help from your teeth and tongue. Some mechanical breakdown continues in your stomach as it churns the food around. During **chemical breakdown**, substances in your digestive system break down food into even smaller particles. Chemical breakdown begins in your mouth, but occurs mostly in your stomach and intestines.

Does it matter if mechanical breakdown occurs? Find out by modeling the process of food breakdown.

CHALLENGE

Why is it important to chew your food?

MATERIALS

For the class

access to watch or wall clock with a second hand

balances (optional)

For each group of four students

- 4 antacid tablets
- 1 120-mL bottle of vinegar
- 2 SEPUP trays
- 2 30-mL graduated cups

SAFETY

This activity uses vinegar, which has a strong odor.

PROCEDURE

Part One: Testing the Model

The Model	
Material/Process	**Represents**
Antacid tablet	Food
Breaking the tablet	Mechanical breakdown
Adding vinegar	Chemical breakdown

1. Model mechanical breakdown by breaking one antacid tablet into four equal-sized pieces. Imagine that each piece is a small piece of food. Place one piece of food into Cup A of a SEPUP tray.
2. Measure 5 mL of vinegar into a 30-mL cup.

3. Model chemical breakdown by adding the vinegar to Cup A. Observe the reaction until it is over. Then record your observations in your science notebook.

4. Discuss your observations with your group.

Part Two: Designing the Experiment

5. Review the Materials list. With your group, design an experiment to test the effect of mechanical breakdown on the speed of chemical breakdown. In other words, how does the size of your food affect the speed at which chemical breakdown occurs?

   When designing your experiment, think about these questions:

   - What is the purpose of your experiment?
   - What variable are you testing?
   - What is your hypothesis?
   - What variables will you keep the same?
   - What is your control?
   - How many trials will you conduct?
   - Will you collect qualitative and/or quantitative data? How will these data help you form a conclusion?
   - How will you record these data?

6. Record your hypothesis and your planned experimental procedure in your science notebook.

7. Make a data table that has space for all the data you need to record. You will fill it in during your experiment.

8. Obtain your teacher's approval of your experiment.

9. Conduct your experiment and record your results.

10. Create a bar graph of your data. Be sure to label your axes and title your graph.

11. If you have time and additional materials are available, revise your procedure and repeat your experiment.

ANALYSIS

1. **a.** In your experiment, what variables did you keep the same?

   b. Were there any variables (except for the one being tested) that you could not keep the same?

   c. How could you or did you improve the design of your experiment? Explain.

2. **a.** What part of digestion was modeled by breaking the tablet?

   b. What part of digestion was modeled by adding vinegar?

3. How does the size of your food affect the speed at which chemical breakdown occurs? Explain how your conclusions are based on the data collected during your experiment, and whether your hypothesis was supported or disproved.

4. Were your conclusions based on qualitative or quantitative data? Explain.

5. Besides preventing choking, why is it important to chew your food?

15 Digestion: An Absorbing Tale

READING

Your digestive system is responsible for both mechanical and chemical breakdown. Everything you eat and drink, including medicines, enters your body through this system. What happens as these substances travel through your body?

CHALLENGE

How does your digestive system work?

MATERIALS

For each student

1 **Student Sheet 15.1, "Your Digestive System"**

READING

Food Breakdown

Take a moment to look at the diagram of the digestive system in Figure 1. You can think of your digestive system as a long tube that goes through your body. Food is absorbed along this tube. If your body didn't absorb what it needed from the food you eat, everything you swallow would come out the other end! You know that doesn't happen. But do you know why? What are the functions of each part of your digestive system?

The digestive system breaks down food into forms that the body can absorb. This breakdown occurs two ways—mechanically and chemically—and it begins in your mouth. Your teeth begin the process of mechanical breakdown. Chemicals in your saliva begin the process of chemical breakdown.

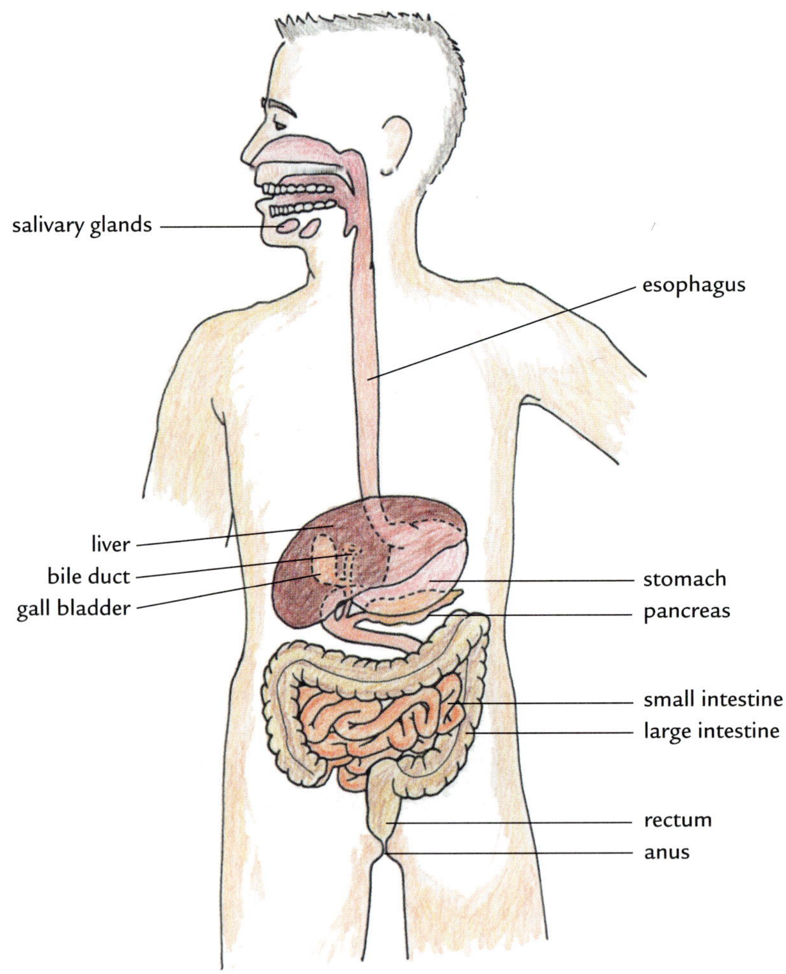

Figure 1: The Human Digestive System

As you swallow, food travels down through your esophagus (ih-SAW-fuh-gus), which is a tube surrounded by muscle. This muscle contracts to help food reach your stomach, a large bag-like organ. Muscles in your stomach wall help to mix the stomach contents. This continues the process of mechanical breakdown. In your stomach, hydrochloric (hi-druh-KLOR-ik) acid—one of the acids used in science laboratories and industries—and other chemicals continue the chemical breakdown of food. The hydrochloric acid in your stomach is so powerful that your stomach is lined with mucus to protect itself. When this lining is absent, ulcers (sores in the lining of the stomach) can form. A high level of hydrochloric acid causes the burning sensation you may feel when you vomit or have indigestion.

STOPPING TO THINK 1

a. How does your mouth contribute to the process of digestion?

b. Explain how your stomach helps break down food.

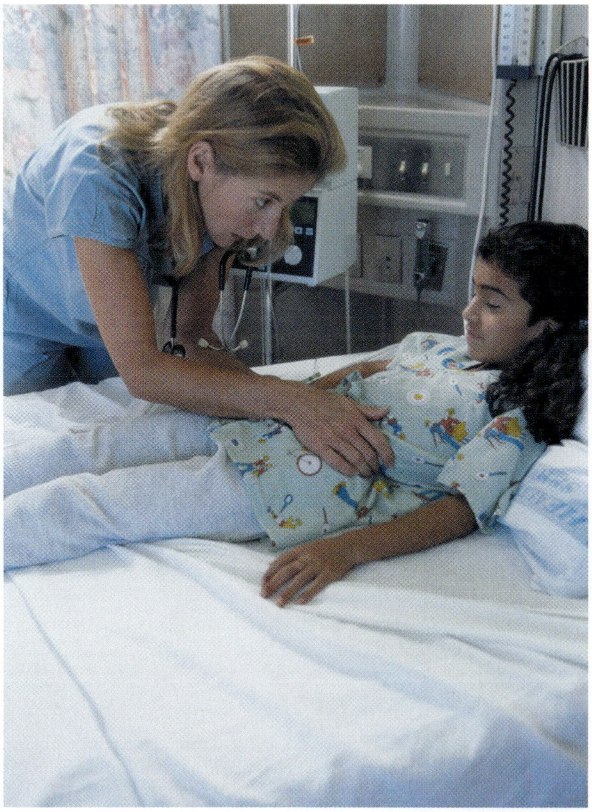

A doctor who specializes in stomach and intestinal problems is called a gastroenterologist (GAS-tro-en-tuh-RAH-luh-jist).

By the time food reaches your small intestine, you wouldn't recognize it anymore! It is a thick pasty mixture. Your small intestine then completes the process of chemical breakdown with help from your pancreas (PAN-kree-us) and liver. As food comes into your small intestine from your stomach, it contains high levels of acid. Your pancreas produces a chemical that reduces this acid level. It also produces chemicals that help break down the proteins and fat found in food. Your liver produces bile, an important mixture that helps break down fat. All of these chemicals combine with the partly broken down food as it travels down your small intestine.

Absorption of Nutrients

Another important process happens in your small intestine, where most of the substances produced by the breakdown of food are absorbed into your blood. After food is completely broken down, we call the pieces **nutrients** (NEW-tree-unts). In the process of **absorption** (ub-SORP-shun), nutrients leave your digestive system and move into your blood, which carries nutrients to the rest of your body. Nutrients are required by all the parts of the body, not just the stomach. The blood acts as the transport vehicle after the stomach has digested food, producing nutrients for all parts of the body.

STOPPING TO THINK 2

a. Explain the relationship between food and nutrients.

b. What role(s) does your small intestine play in digestion?

The fact that most of the final breakdown and nutrient absorption occurs in your small intestine may help explain its length. The average adult small intestine is 5–6 meters (about 15–18 feet) long! This length, plus the folds in the wall of the small intestine (see Figure 2),

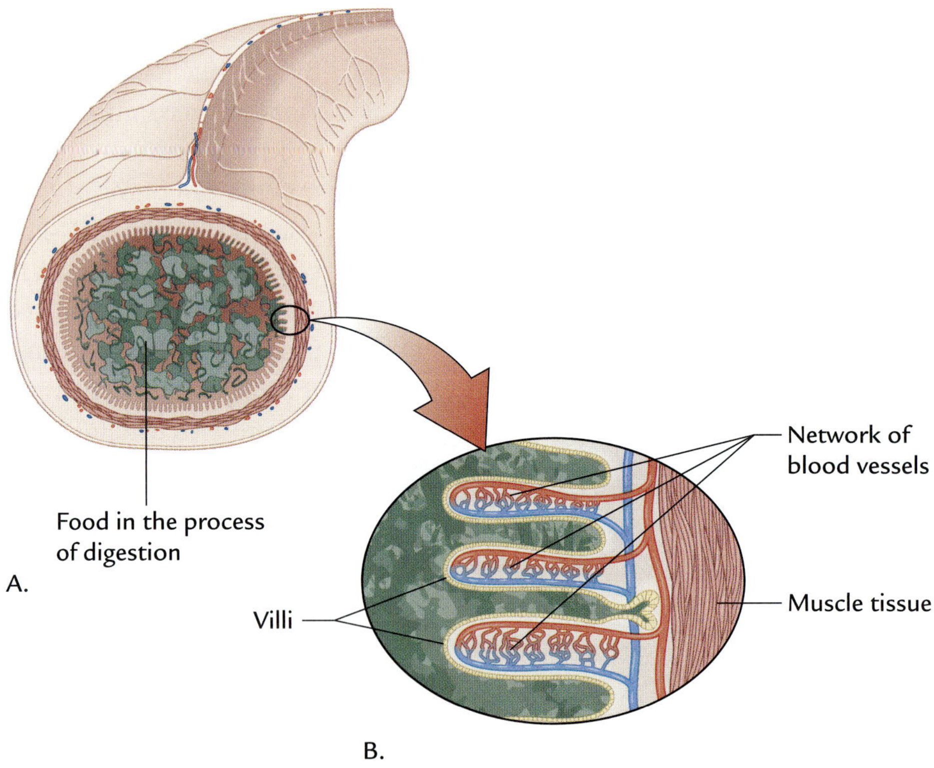

Figure 2: Cross-Section of Your Small Intestine
Nutrients are absorbed by your blood across the wall of your small intestine. Fingerlike projections from the wall of your small intestine are known as villi (VIL-eye) (singular, villus*). Nutrients must pass through villi and the walls of tiny blood vessels to enter your blood.*

provides lots of surface area for nutrient absorption. Your blood transports these nutrients to different parts of your body, but first it makes an important stop.

All of the blood that leaves your stomach and intestines goes directly to your liver before traveling throughout the rest of your body. This is because your liver performs two important functions related to digestion, besides producing bile. First, it breaks down toxins such as alcohol and some medicines. (Your blood is later filtered by your kidneys, which excrete liquid wastes and some dissolved toxins as urine.) Second, it processes nutrients into forms that are easier for the rest of your body to use. For example, your liver stores carbohydrates. When you suddenly need energy, it converts these carbohydrates to sugars that your body can use.

STOPPING TO THINK 3

Why does blood travel to your liver before transporting nutrients to other parts of your body?

Getting Rid of Solid Waste

Any material that has not been absorbed by your small intestine continues down into your large intestine, or colon (KOLE-un). In your large intestine, large quantities of water and some remaining vitamins are absorbed into your blood. The remaining unabsorbed material forms a solid waste as it travels through the large intestine, a process that can take 18–24 hours. This solid waste is stored in the rectum (REK-tum) before being pushed out through the anus (AY-nus). What is this solid waste made of? It contains bacteria, substances that your body can't digest, and some remaining water. Bacteria live and grow in your intestines, and they help you in several ways. They break down some plant material that your body can't break down on its own, they make vitamin K, and they help prevent harmful bacteria from finding a home. The trade-off for providing a home for these helpful bacteria is the gas and odors they produce.

STOPPING TO THINK 4

The reading describes three components of human solid waste. Which two of these do you think are the main components?

ANALYSIS

1. What are some of the functions of the digestive system?

2. Copy the table below. Then fill in the table by placing an "X" to indicate the function(s) of each organ. The first row has been done for you.

Functions of Digestive Organs

Organ (or Structure)	Mechanical Breakdown	Chemical Breakdown	Nutrient Absorption	Water Absorption and Solid Waste Production
Mouth	X	X		
Stomach				
Small intestine				
Pancreas				
Liver				
Large intestine				

3. Imagine taking a bite of a burrito. Follow the beans in the burrito through the process of digestion. Explain what types of changes take place and where each change happens.

4. Most substances are absorbed in the small intestine and not in the stomach. Aspirin is a common exception; it is absorbed in the stomach. Some alcohol is absorbed in the stomach, but most is absorbed in the intestine.

   a. Why would you want medicines, like aspirin, to be absorbed in the stomach instead of the small intestine?

   b. What is the effect of some alcohol being absorbed in the stomach?

5. Take a closer look at the villi of the small intestine (shown in Figure 2b). How do the villi help nutrients move into the blood quickly? **Hint:** What would happen if there were no villi, only a smooth surface?

16 Balancing Act

How does your body use the nutrients it absorbs? Even if you don't exercise, you use (or "burn") energy all the time. Just sitting in a chair for 30 minutes can consume up to 30 Calories! That is why the Food and Drug Administration (FDA) recommends that most kids eat about 2,200 Calories each day. You get these Calories from the food you eat.

Food is made up mostly of carbohydrates, fats, and proteins. One important role for carbohydrates in the body is to provide quick energy. That's why distance runners and other endurance athletes often eat carbohydrates just before a race ("carbo-loading"). Fats, on the other hand, store energy over long periods. Proteins may be used for building muscle and other tissues. They may also be burned for energy when carbohydrates and fats are not available.

How many Calories do *you* need for energy? Consider the energy equation in Figure 1.

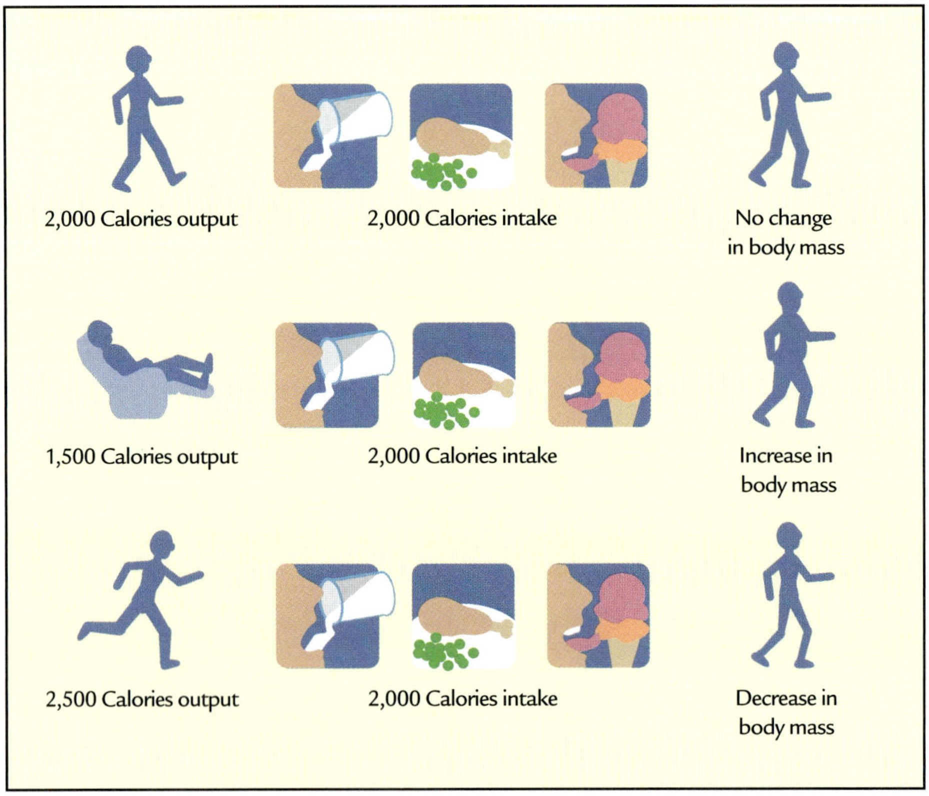

Figure 1: The Energy Equation

CHALLENGE

Can you design and make an energy bar that

- provides between 150–250 Calories?
- provides *no more* than 75 Calories from fat?
- provides *no less* than 25 Calories from protein?
- costs less than $0.50 to make?
- tastes good?
- resists crushing under the weight of a textbook?

MATERIALS

For each group of four

	various ingredients with nutritional information labels
1	set of measuring spoons
1	¼-cup measuring cup
1	textbook

For each pair of students

1	paper plate
1	plastic spoon
1	plastic knife
1–2	plastic sandwich bags
1	calculator

For each student

1	Student Sheet 16.1, "Energy Bar Data and Calculations"

SAFETY

Never taste materials or eat or drink in science class unless specifically told to do so by your teacher. Be sure that your work area is clean and free of any materials not needed for this activity. If you are allergic to peanuts or dairy products, or if you have any other health issue, such as diabetes, that limits what you can eat, tell your teacher and do not taste the energy bars in this activity.

Wash your hands before and after preparing the energy bars, and use only food-grade plates and utensils. Do not contaminate foods by licking your fingers or the utensils.

PROCEDURE

Part 1: Designing the Energy Bar

1. Read the energy bar requirements described in the Challenge. Then look at the information provided in Table 1, "Data on Ingredients," on Student Sheet 16.1, "Energy Bar Data and Calculations."
2. Discuss with your group what ingredients you could use to make an energy bar that would meet all of the requirements. Be sure to consider ingredients that are not listed on Student Sheet 16.1.
3. Decide who will bring in the additional ingredients, and gather your materials.
4. Record any additional ingredients you are using in Table 1 of Student Sheet 16.1. Use the nutritional label on the package to fill in the information for each ingredient in Table 1.
5. With your partner, write a recipe for making an energy bar. Include how much of each ingredient you plan to use. **Hint:** Try to use even amounts, such as ¼ cup or 2 tablespoons, to make your calculations easier.
6. Record the ingredients you are using and the amounts of each one in Table 2, "Calculations for My Energy Bar," of Student Sheet 16.1. Be sure to include the units of the amounts you plan to use.

7. Use Table 2 to calculate the number of servings, calories, and cost for each ingredient.

8. Use Table 3, "My Energy Bar Totals," to find out if your energy bar would meet the calorie and cost requirements described in the Challenge.

Part 2: Testing the Energy Bar

9. Follow your recipe and make your first energy bar. Place the bar in a plastic bag and label it with your name and "Energy Bar 1."

10. As directed by your teacher, test the bar for resistance to crushing. Record the results in Table 4, "Evaluating My Energy Bar," on Student Sheet 16.1.

11. Break the bar in half. Leave one half in the bag with the label. Taste the other half and rate it using the scale shown on the left. Record the results of your taste test on Student Sheet 16.1.

Taste Scale	
Good	4
OK	3
Edible	2
Terrible	1

12. Use the rest of your data on Energy Bar 1 to complete Table 4 on Student Sheet 16.1. Then discuss with your group how you could improve your energy bar recipe.

13. If you have the time and materials, revise your energy bar recipe to make and test a second energy bar.

ANALYSIS

1. What are the similarities between designing a product, like the energy bar, and conducting a scientific investigation? What are the differences?

2. Which requirements did your energy bar meet? Were there any requirements that you could not meet? Explain.

3. The United States Department of Agriculture (USDA) recommends that a person eating a healthy diet consume less than 30% of his or her daily Calories from fat. For a person eating 2,200 Calories every day, this means no more than 660 Calories from fat.

   a. Copy and complete a table like the one shown below.

Energy Bar Calories and Fat

	Calories in My Energy Bar	Calories USDA Recommends for One Day	My Bar's Percentage of USDA Recommendation
Total Calories			
Calories from fat			

   b. How many of your energy bars could a person eat in one day without exceeding 2,200 Calories, assuming he or she ate nothing else?

   c. How many of your energy bars could a person eat in one day without exceeding 660 Calories from fat, assuming that he or she ate nothing else?

4. Explain whether you would recommend your energy bar as a healthy snack. Support your answer with *quantitative* evidence and identify the trade-offs of your decision.

   Hint: To write a complete answer, first state your opinion. Provide two or more pieces of evidence that support your opinion. Then consider all sides of the issue and identify the trade-offs of your decision.

5. Look again at Figure 1, "The Energy Equation." Write an explanation of how this figure helps to explain changes in body weight in terms of a balance between food (energy input) and exercise (energy use).

EXTENSIONS

Table 1: Burning Calories	
Activity	**Calories**
Basketball	7
Biking (12–14 mph)	7
Dancing (fast)	5
Football	7–8
Gymnastics	4
Rollerblading	6
Running (6 mph)	9
Sitting (in class)	2
Skateboarding	4
Sleeping	1
Soccer	6
Softball	4
Standing	1
Swimming laps	9
Tennis	6
Walking (4 mph)	4
Watching T.V.	1

Approximate number of calories consumed during 1 minute of activity by a 110-pound person.

1. Imagine that you are in charge of marketing your energy bar. Design an ad to sell it. What further testing and changes would you suggest before your company starts manufacturing your energy bar?

2. *Is your Calorie intake balanced by your level of activity?* Choose a food label from one of your favorite snack foods. How many Calories are provided by one serving of this food? How many servings of this food do you usually eat at one time? Multiply your answers to figure out how many Calories you usually consume when you eat this food.

Look at Table 1, "Burning Calories." Create a weekly activity table to record which activities you do and the number of minutes you do each of them each week. Use Table 1 (or go to the SALI page of the SEPUP website) to calculate approximately how many Calories you burn in an average week.

Choose one of your favorite activities from your weekly activity table. How long would you have to do this activity to use up the number of Calories you usually consume when you eat your favorite snack food (calculated above)?

17 Gas Exchange

LABORATORY

Your blood transports the nutrients that you eat to different parts of your body. It also carries oxygen from your lungs to other organs and tissues. With every breath you take, you inhale oxygen and exhale carbon dioxide. Your body uses the oxygen to get energy from food. When your body breaks down food, it produces wastes. One of the wastes is carbon dioxide.

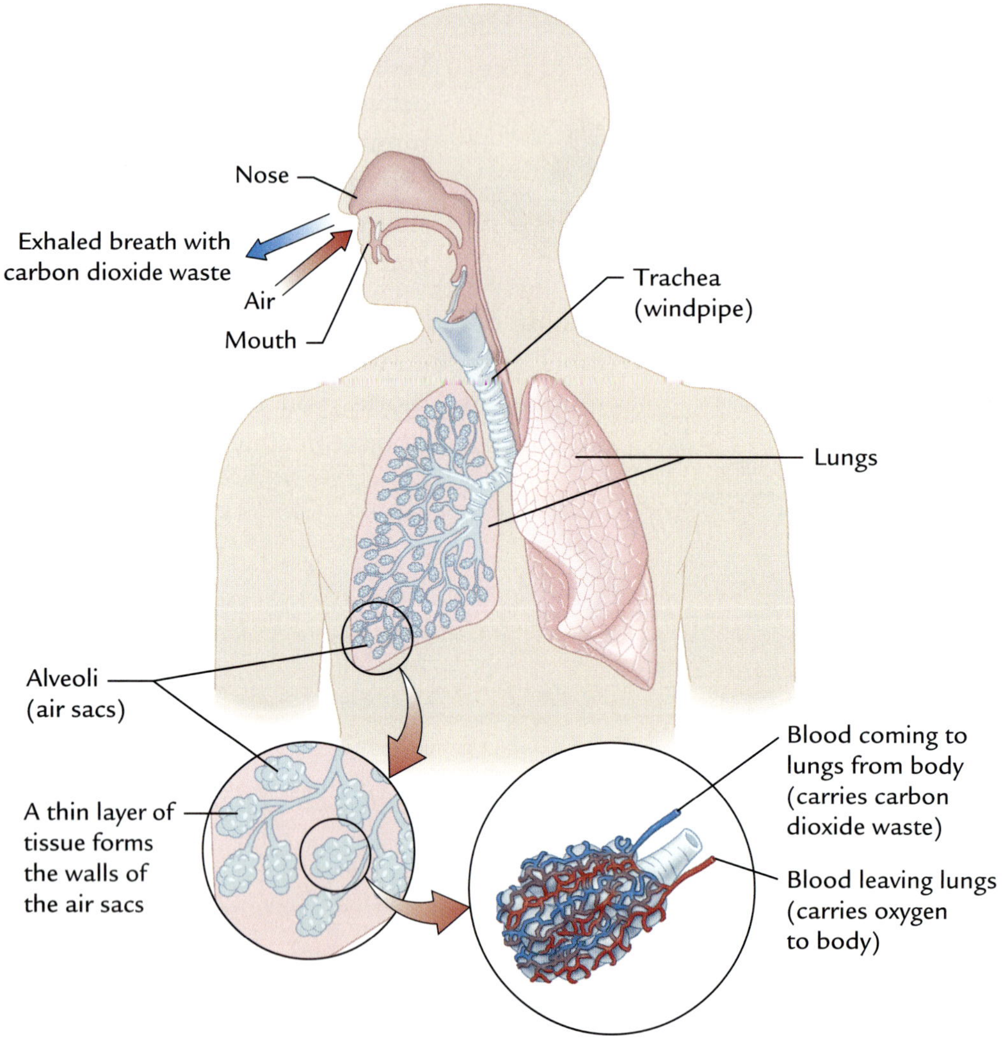

Figure 1: Human Respiratory System

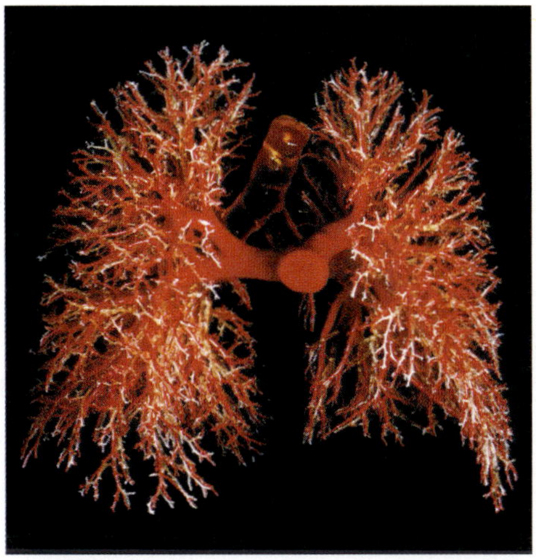

This is a plastic mold of the air ways and air sacs in the lungs. Compare it to the diagram of the lungs on the opposite page.

Indicators (IN-duh-kay-ters) are chemicals that change their appearance in different types of solutions. You will work with the indicator bromthymol (brome-THY-mall) blue, also known as BTB. BTB can be either blue or yellow. When added to a solution containing carbon dioxide, BTB is yellow.

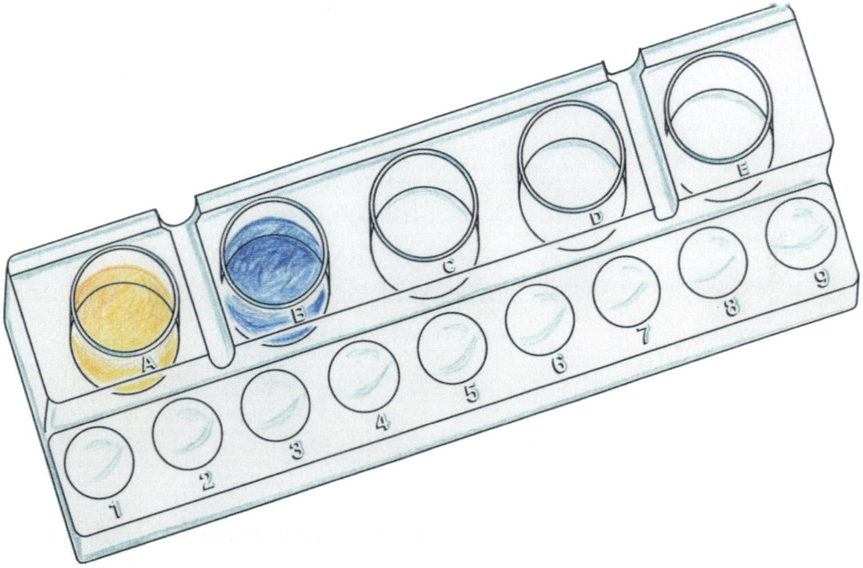

Both solutions contain the indicator BTB. Which cup has a solution containing carbon dioxide?

CHALLENGE

How much carbon dioxide is in your exhaled breath?

MATERIALS

For each group of four students

1 dropper bottle of bromthymol blue (BTB) indicator
5 plastic cups
supply of water

For each pair of students

1 dropper bottle of 0.05 M sodium hydroxide
1 SEPUP tray
1 dropper
1 30-mL graduated cup

For each student

1 1-gallon plastic bag
1 straw
1 stir stick
access to a wall clock or watch with a second hand

SAFETY

In this activity, you will be blowing through a straw into chemicals. Do not inhale through the straw! Breathe in through your nose and exhale through your mouth. If you accidentally swallow liquid, rinse your mouth thoroughly and drink plenty of water. Be sure to tell your teacher.

PROCEDURE

Part One: Using BTB to Test for Carbon Dioxide

1. Work with your partner to add 5 mL water to each of the five large cups (A–E) of your SEPUP tray. Use the 30-mL graduated cup to measure the water.
2. Add 2 drops of BTB to each cup and stir.
3. Create a data table to record the initial and final colors of the solutions in each cup. Record the initial colors now. Cup A will provide a control.

4. Use your dropper to bubble air into Cup B. Place the dropper into the solution and press the air out of the bulb. Before releasing the bulb, remove the tip from the solution. This will prevent uptake of solution into the dropper. (If you accidentally get solution into the dropper, simply squirt it back into Cup B.) Repeat this for 15 seconds.

5. Record the final color of the solution in Cup B in your data table.

6. Add 3 drops of 0.05 M sodium hydroxide to Cup C. Record the final color in your data table.

7. Unwrap your straw and place one end in Cup D. Take a deep breath, and then gently blow through the straw for 15 seconds. (Remember not to inhale through the straw!) Record the final color of the solution in Cup D in your data table.

8. Have your partner blow through a clean straw into Cup E for 15 seconds. (Remember not to inhale through the straw!) Record the final color in your data table.

9. Add 3 drops of sodium hydroxide to Cups D and E. In your science notebook, record any changes that you observe.

10. Work with your partner to complete Analysis Questions 1 and 2.

Part Two: Using BTB to Measure Carbon Dioxide in Exhaled Breath

11. Work with another pair of students to set up a control:

    a. Measure 10 mL of water using the 30-mL graduated cup.

    b. Add 3 drops of BTB to the graduated cup and stir.

    c. Pour the BTB solution into a large plastic cup. This solution will be the control for every member of your group.

12. Have each person in your group set up his or her own bag of BTB solution:

    a. Measure 10 mL of water using the 30-mL graduated cup.

    b. Add 3 drops of BTB to the graduated cup and stir.

    c. Pour the BTB solution into your own 1-gallon plastic bag.

13. Remove the air from your plastic bag by slowly flattening it. Be careful not to spill any of the BTB solution out of the bag. While keeping the air out of the bag, place a straw in the mouth of the bag. Make an air-tight seal by holding the mouth of the bag tightly around the straw.

14. Be sure you are sitting down. Then fill the bag with air from your lungs by blowing through the straw until the bag is fully inflated. When you finish blowing, pull out the straw. As you pull out the straw, squeeze the bag tightly shut so no air escapes.

15. Holding the bag closed, shake the bag vigorously 25 times.

16. Pour the BTB solution from the bag into a clean, empty plastic cup.

17. *How much carbon dioxide is in your exhaled breath?* You can find out by counting how many drops of sodium hydroxide are needed to make your BTB solution the same color as the control:

a. Add 1 drop of sodium hydroxide to your plastic cup.

b. Gently stir the solution and wait at least 10 seconds.

c. Record in your science notebook that you added 1 drop.

d. Compare the color of your solution to the control. *Is it the same color as the control for at least 30 seconds?*

If your answer is no, repeat Steps 17a–d. Be sure to keep track of the total number of drops!

If your answer is yes, go on to Step 18.

18. In your science notebook, record the total number of drops it took to change your solution back to the same color as the control. Then record your total on the class data table.

19. Draw a bar graph with the class results. Remember to title your graph and label the axes!

EXTENSION 1

Do you exhale more carbon dioxide after you hold your breath? Find out by modifying and repeating Part Two of the Procedure.

ANALYSIS

Part One: Using BTB to Test for Carbon Dioxide

1. What was the purpose of the solution in Cup A?

2. a. Which of the solutions in Part One contained carbon dioxide? Support your answer with evidence from your experimental results.

   b. What does this tell you about the exhaled breath of human beings?

Part Two: Using BTB to Measure Carbon Dioxide in Exhaled Breath

3. Review the class data table. What was the range of carbon dioxide in exhaled breath (as measured by drops of sodium hydroxide)? **Hint:** You looked at the range of class data in Activity 9, "Data Toss."

4. Look again at Figure 1, "Human Respiratory System." Considering all the oxygen that has to get into your blood and all the carbon dioxide that has to escape from your blood, why do you think the inside of the lung is structured the way it is?

5. **a.** Were the data collected in Part One qualitative or quantitative? Explain.

   b. Were the data collected in Part Two qualitative or quantitative? Explain.

6. **a.** Look carefully at Figure 1, "Human Respiratory System." What are some of the important structures in the respiratory system?

   b. Explain how gases are exchanged within the respiratory system.

EXTENSION 2

How do you think your body gets more oxygen when you exercise? Do you breathe faster (take more breaths per minute)? Or do you absorb more oxygen from the air with each breath? Use what you learned in this activity to develop an experiment to test your hypothesis.

18 The Circulation Game

MODELING

How do the nutrients absorbed from the digestive system get to every part of your body? They travel in your blood through your **circulatory** (SIR-kyu-luh-tor-ee) **system**. In Figure 1, you can see that your blood goes to every part of your body. The main function of your blood is to transport, or carry things around. Blood transports oxygen, nutrients, and wastes such as carbon dioxide. Some organs in your body help get nutrients and oxygen into your blood and also remove wastes. In this activity, you will model the path of your blood as it moves through your body and you will learn what happens along the way.

CHALLENGE

What does blood do as it travels around your body?

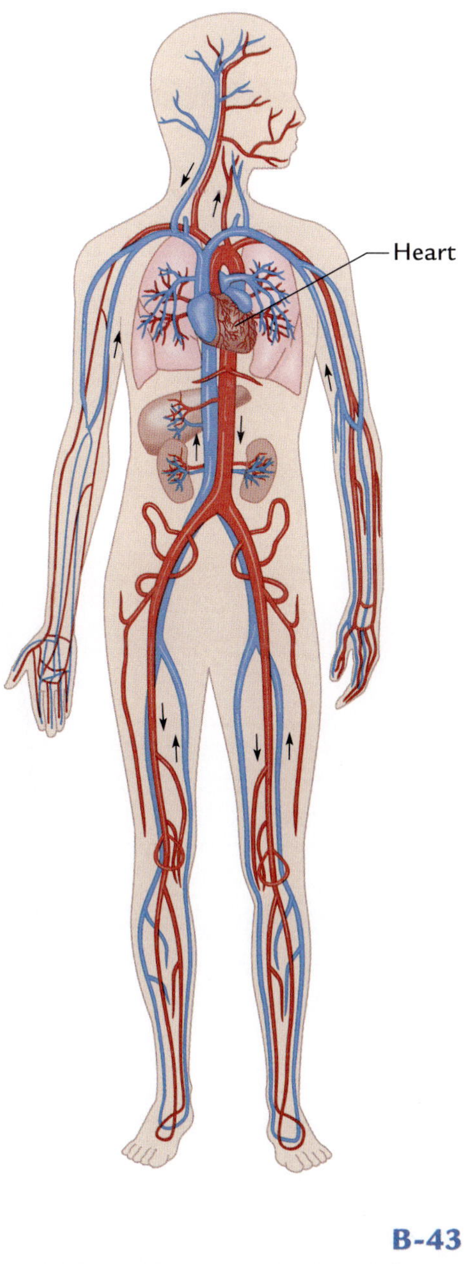

Figure 1: Human Circulatory System

MATERIALS

Part Two: Modeling Circulation

For the class

1 set of ***oxygen***, ***nutrient***, ***carbon dioxide***, and ***waste*** cards

For each student

1 Role Card

1 paper clip, safety pin, hole-punch and string, or piece of transparent tape

PROCEDURE

Part One: Blood Flow

1. Look at Figure 2, "Diagram of Blood Flow." This is a simplified map of how blood travels around your body. Use your finger to trace one of the possible paths of blood flow. Begin on the left side of the heart (on your right) and stop once you reach the left side of the heart again. Be sure to go in the direction of the arrows.

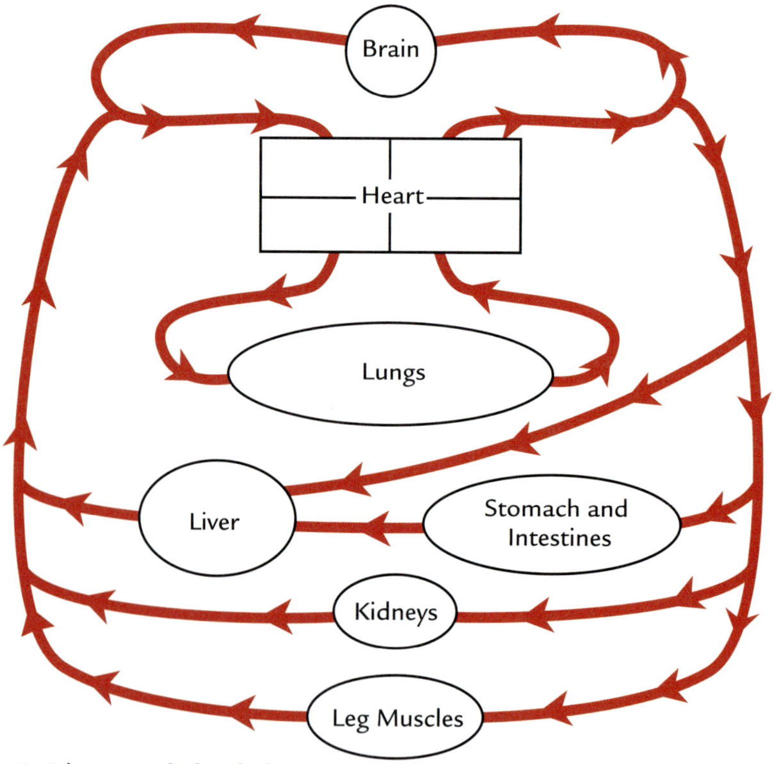

Figure 2: Diagram of Blood Flow

2. In your science notebook, record which organs and structures you passed through in your path.

3. Repeat Steps 1 and 2 by tracing a different path of blood through the human body.

4. Discuss the following questions with your group members:

   - Did everyone trace the same paths? If not, compare the organs (and structures) along the different paths.

   - Which organs does the blood have to pass through each time it goes around the human body?

   - Why do you think blood always has to pass through these organs?

5. Use your discussion and your knowledge of the human body to complete a table like Table 1.

Table 1: Functions of Certain Organs

Function	Organ(s)
Pumps blood	
Brings oxygen into the body	
Carries carbon dioxide out of the body	
Absorbs nutrients	
Removes wastes	

6. As directed by your teacher, share your discussion with the rest of the class.

Part Two: Modeling Circulation

THE MODEL

You will model the way in which your blood transports oxygen, nutrients, and carbon dioxide and other wastes. Each student will receive a Role Card to role play one part of the human body: blood, brain, heart, lungs, stomach and intestines, liver, kidneys, or leg muscles.

In this model, as "blood" flows through the "human body," it will absorb "oxygen" and "nutrients" and carry them to other parts of the body. "Organs" will use the nutrients and oxygen and get rid of carbon dioxide and other wastes by giving them to the blood. Colored cards will represent the four substances (oxygen, nutrients, carbon dioxide, and waste) that are being transported.

7. Your teacher will give you a Role Card. Read your Role Card carefully to see what your job will be. Be sure to note how many *oxygen, nutrient, carbon dioxide,* and *waste* cards you will need to collect before beginning.
8. Collect the cards that you need to begin. Keep *oxygen, nutrient, carbon dioxide,* and *waste* cards in separate stacks so that exchanges can take place quickly. Then attach your Role Card to your clothing.
9. Find your place on the blood flow diagram before you begin the game.
10. As a class, play the Circulation Game.
11. Switch Role Cards before doing more trials.

ANALYSIS

1. Compare this circulation model to the human body. How well did the Circulation Game represent what really happens inside your body?

2. Do all parts of the human body use oxygen and nutrients? Explain your answer.

3. Why does blood flow from the stomach and intestines directly to the liver? **Hint:** Review your notes from Activity 13, "Living with Your Liver," and Activity 15, "Digestion—An Absorbing Tale," for help.

4. What are the functions of the blood as it travels around the human body? Be specific.

5. Look at the diagram below.

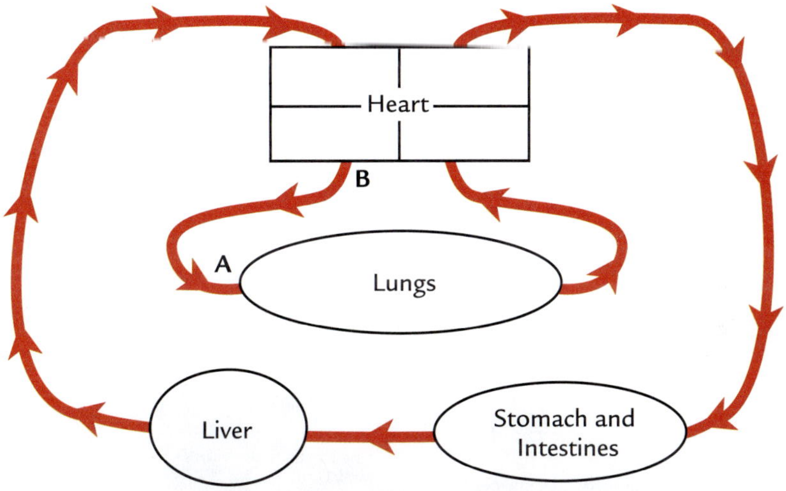

   a. Use your finger to trace the path between Point A and Point B, making sure to follow the direction of the arrows. List the organs in the order in which you passed through them.

   b. Imagine blood carrying only carbon dioxide and nutrients at Point A. Describe what happens to the blood as it flows from Point A to Point B.

19 Heart-ily Fit

INVESTIGATION

Imagine going on a long hike. How would this exercise affect your body? How could you measure this effect? When you exercise, your heart beats faster to provide more oxygen to the muscles in your body. The harder you exercise, the faster your heart beats. You can measure the speed at which your heart beats by taking your **pulse**, which reflects the contractions of your heart. Each time your heart beats, it sends blood through your body. You can feel this surge when you press on blood vessels near the surface of your skin.

One way you can investigate your level of fitness is by measuring how long it takes for your heart to recover from exercise. The more physically fit a person is, the more efficiently oxygen and nutrients can be transported to the muscles. The faster your heart recovers from exercise, the more fit you are.

CHALLENGE

How can you quantitatively measure your level of fitness?

Did you know that you can have a career studying exercise and helping people get into shape and stay fit? This type of work is known as exercise physiology (fih-zee-AH-luh-jee).

MATERIALS

For the class

access to a wall clock or watch with a second hand

For each pair of students

1 calculator

For each student

1 Student Sheet 19.1, "Pulse Data"

SAFETY

If you begin to feel dizzy or short of breath during exercise, stop exercising immediately. Do not participate in this activity if you have any condition that prevents you from exercising.

PROCEDURE

Part One: Resting Pulse

You can monitor your pulse at a number of sites. Two convenient sites to use are:

- *the artery at the base of the wrist (of either hand)*
- *the artery at the side of the neck.*

1. To take your pulse at rest, sit comfortably. (The best and most accurate time to measure resting pulse is immediately after you wake up in the morning, before you get out of bed. You can check your resting pulse at that time to see how it varies from the values taken in class.)

2. Use the first two fingers of one hand to locate your pulse at the base of your wrist.

3. Measure your pulse for 15 seconds. Have your partner keep track of the time.

4. Record your data for Trial 1 on Student Sheet 19.1, "Pulse Data."

5. Repeat Step 3 two more times and record your data for Trials 2 and 3.

6. Repeat Steps 1–5 for your partner.

7. Calculate your pulse for 60 seconds (1 minute) by multiplying each 15-second pulse by 4.

8. Calculate your average resting pulse per minute. Do this by adding all the numbers in the column titled "60-Second Pulse." Then divide your total by 3. Record your average resting pulse per minute on Student Sheet 19.1.

Part Two: Recovery Time

9. As discussed in class, exercise for 5 minutes. You should begin to feel your heart beating faster. **Hint:** If you can have a normal conversation while exercising, you are not exercising hard enough. If you cannot talk at all, you are exercising too hard.

10. After 5 minutes, stop exercising and sit down. Immediately begin taking your pulse every 30 seconds for the next 5 minutes. Record your 15-second pulse every 30 seconds on Student Sheet 19.1.

11. Calculate your 60-second pulse for each time period by multiplying each 15-second pulse by 4.

ANALYSIS

1. What happened to your breathing rate during exercise? Discuss what was happening inside your body that caused this to happen.

2. What caused the difference between your resting pulse and your pulse after exercise? In other words, what was happening inside your body that caused your pulse to change?

3. a. *Recovery time* is the time it takes for your pulse to return to within 20% of your resting pulse. In order to measure your recovery time, you must first know when you are within 20% of your resting pulse. Calculate this value by multiplying your resting pulse by 1.2.

Resting pulse X 1.2 = _______ beats/min

b. Look at your 60-second pulse values in Section III, "Recovery Time," of Student Sheet 19.1. How many minutes after you stopped exercising did it take you to return to within 20% of your resting pulse? (This is your recovery time.)

4. Use the guide below to create a line graph of pulse vs. time. Plot the data from your 60-second pulse during the recovery period.

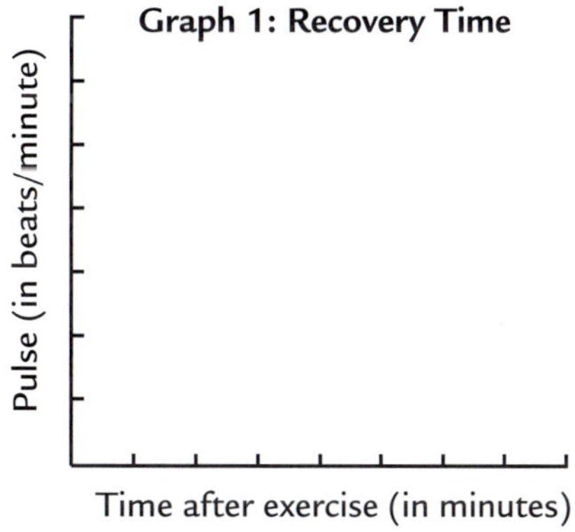

5. If you improved your level of physical fitness, would you expect your resting pulse to increase or decrease? Explain.

6. What do you predict would happen to your recovery time if you exercised at least three times a week for a month?

EXTENSION

What exercises do you enjoy? Identify an exercise that you could practice to improve your recovery time. Or go to the SALI page of the SEPUP website for some ideas. Then develop a plan to exercise regularly for the next month. You should exercise at least three times a week for at least 20 minutes each time. After exercising for a month, repeat Part Two of the Procedure. What effect does regular exercise have on your recovery time?

20 Great-Aunt Lily's Will

Have you ever heard of the American Heart Association, the American Cancer Society, or the American Lung Association? Each of these organizations raises money for research to fight specific diseases. For example, the American Heart Association focuses on diseases of the cardiovascular (CAR-dee-oh-VAS-kyu-lur) system, which you know as the circulatory system. People at the Association must decide how to spend the money they raise to fight heart disease. Sometimes they develop brochures like the ones shown here to help people learn about a disease. These brochures are one way organizations work to improve public health. There are many other ways.

Reproduced with permission from the American Lung Association. "Shop Smart with Heart" ©2001, American Heart Association.

Would you like to develop a public health campaign? Some people do this as a career. Imagine you have to decide how to spend a limited amount of money on public health. What would you do?

CHALLENGE

What are the trade-offs involved in choosing research, treatment, or education as approaches to promote public health?

MATERIALS

For each student

1 Student Sheet 20.1, "Requests for Funding"

PROCEDURE

1. Watch the role play of the story of Great-Aunt Lily's will. As you watch, decide what you would do if you were in Julia's place.
2. Use Table 1, "Awarding Money," on Student Sheet 20.1, "Requests for Funding," to record which organizations have requested money and how much each organization has requested.
3. In Table 1, identify whether the organization has requested money for education, research, or treatment.
4. Complete Table 1 by deciding how much money you will award to each organization. Remember, your total cannot exceed 1 million dollars.
5. Calculate the percent of the 1 million dollars you awarded for education, research, or treatment by completing Table 2, "Percent Awarded," on Student Sheet 20.1.

GREAT-AUNT LILY'S WILL

"I just can't decide what to do," Julia moaned as she looked at the papers all over her desk.

Hearing her complain, James, Trina, and Harold poked their heads around the door. "What's up?" asked Harold.

Julia pointed to all the documents on the desk. "In front of me I supposedly have all the information I need to make a decision. You know Great-Aunt Lily left 1 million dollars to 'help fight heart disease.' That's a lot more complicated than it sounds. First, there are reports that say one way to reduce heart disease is to educate people about heart disease and prevention. Second, researchers want money to study various heart problems.

They hope to develop new and better treatments. Finally, some people need help paying for medical procedures."

"Giving money to lots of good causes—sounds like fun to me!" Trina remarked.

"Well, that's true," Julia sighed. "But 1 million dollars isn't enough for such a big problem. Almost 25% of the population has some type of heart disease. In fact, heart disease has been the top killer in the United States every year since 1900! With heart disease so common, a lot of people need money."

James picked up a letter off the desk. "This looks easy. This organization called Making A Difference helps people who can't pay for surgeries. They want $500,000 to help three families pay for the costs of heart surgery.... How sad—one of the families has a little girl who was born with a hole in her heart. That's been fixed, but she needs more surgery. And in this other family, a dad needs a heart transplant."

"I know," replied Julia. "It seems like Making A Difference could really change the lives of these three families. But funding public education on heart disease could reach thousands of people. Providing people with information could mean that they take better care of themselves. This could reduce their risk of heart disease. More people would stay healthier and the cost of medical care would be reduced. The cost of heart diseases and strokes in 1998 alone was 274.2 BILLION dollars!"

"Whoa, that's a lot of money." Harold began to leaf through some of the folders. "But how does education help? No one pays attention to that stuff. After all, everyone believes it'll never happen to them."

"I don't think that's true," said Trina. "Look at all the people who quit smoking when they realized what it does to their heart and lungs. Many people don't even know the early warning signs for heart disease. They don't know what to do when someone has a heart problem, like a heart attack. But people do pick up and read brochures while waiting to see the doctor."

"Exactly." Julia went on. "Producing a million brochures for health service agencies, like doctors' offices, hospitals, and community health clinics, would cost around $400,000. Sending out brochures could potentially help a million people. It might prevent hundreds of cases of heart disease!"

"I see where you got those numbers," James said as he held up a folder. "A group called HeartSmart has asked for $400,000 to do exactly what you just said."

Harold held up another file. "Here's another organization that wants money for public education. Project Heart needs $685,000 to produce and air 15-second spots on television. Their request doesn't say how often or on what channel, but having informational spots on TV is a cool idea! I bet a lot of future heart problems could be prevented that way. In fact, we'd be helping the next generation of adults."

"Harold, you're not helping!" exclaimed Julia. "I haven't even mentioned research. Scientists around the country are trying to identify the causes of heart disease. Other scientists are researching better treatments for heart problems. Without money, some of this research will end. Who knows which research project might result in the next breakthrough!"

"How much does a research project cost?" Harold asked.

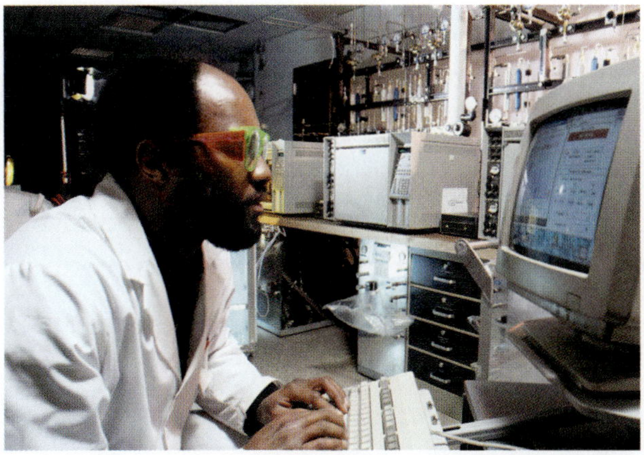

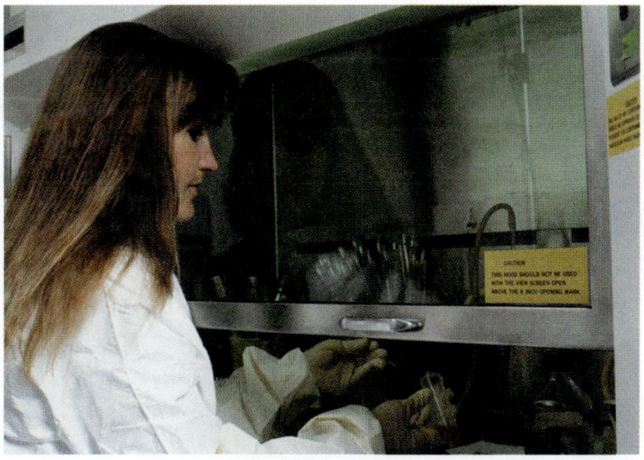

Some research scientists work to develop new drugs or new treatments for disease.

"Research is expensive," replied Trina as she looked at a report. "CardioResearch requested $807,000. They want to continue their research on a drug that would help people with high blood pressure."

James read aloud from another letter. "University Research Hospital would like exactly 1 million dollars to continue its work developing better surgical procedures for blocked blood vessels. Hmmm...that would really help a lot of people whose arteries are blocked. You probably know someone who has had bypass surgery to deal with that problem."

Harold had an idea. "How about splitting up the money? Maybe you could give some of these people only part of what they asked for. Then they might be able to get money from other places as well."

"Thanks, Harold," said Julia. "Now you know exactly where I started: What's the best way to spend the 1 million dollars? Who should get the money? And how much?"

ANALYSIS

1. Discuss with your group why you decided to award the money as you did.
2. What additional information (or types of information) would have helped you make a better decision?
3. What are the trade-offs of choosing only one way—education, research, *or* treatment—to deal with a public health problem such as heart disease?
4. **Reflection:** What effect do you think public education, such as short television spots, has on people's behavior? Support your answer with evidence from your own personal experience.

21 Inside a Pump

LABORATORY

You know that your blood transports oxygen and carbon dioxide from one part of your body to another. Your heart is the pump that keeps your blood moving. Your heart pumps a lot like a mechanical pump. You will find out more about the structure and function of your heart by using a mechanical pump as a model.

CHALLENGE

What type of pump is better for pumping water? What does this tell you about the structure of your heart?

MATERIALS

For the class

- 4 **siphon bulbs with tubing**

For each group of four students

- 1 **pressure bulb**
- 2 **45-cm (18-inch) lengths of clear plastic tubing**
- 2 **plastic bins (or buckets)**
- **supply of water**

SAFETY

Do not attempt to use your mouth to start a pump. During this activity, water may spill on the floor, so walk carefully.

PROCEDURE

1. Fill one of the bins ¾ full of water. (Your teacher may have already done this.)
2. Connect two pieces of plastic tubing to the pressure bulb, one on each end.

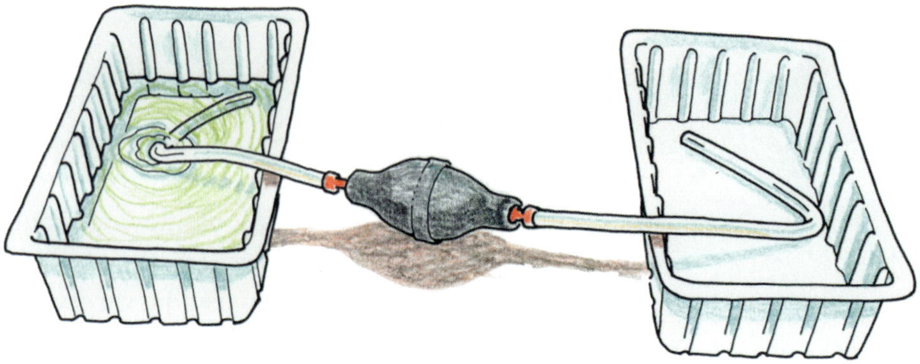

Figure 1: The Pump Set-Up

3. Place one end of the tubing in the container of water and the other in the empty container, as shown in Figure 1.
4. Use your pump (pressure bulb) to transfer the water from one container to the other. (You may have to hold the bulb at or below the level of the water to get it started. Or you can squeeze the bulb and place your finger over the open end of the tubing while releasing the bulb to allow water to flow into the bulb.)
5. Borrow a siphon bulb from your teacher and try to create a pump that can transfer water from one container to the other.
6. Record how you made one or both pumps work. You can draw pictures to show what you did.
7. Try to figure out how the pressure bulb works. **Hint:** You can remove the tubing, shake the bulb, and listen to both ends as you squeeze air through the bulb.
8. Watch your teacher demonstrate how a siphon bulb works.

ANALYSIS

1. If you cut lengthwise through each bulb, the outer walls of the bulbs will look like the diagrams below. These are called lengthwise *cross-sections*. A cross-section gives a view inside a sliced object.

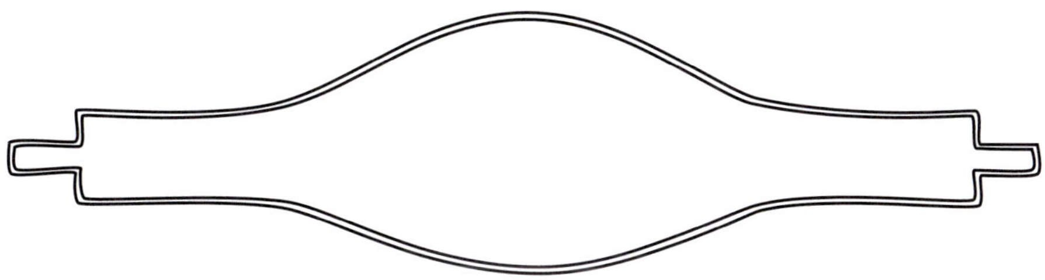

Pressure bulb

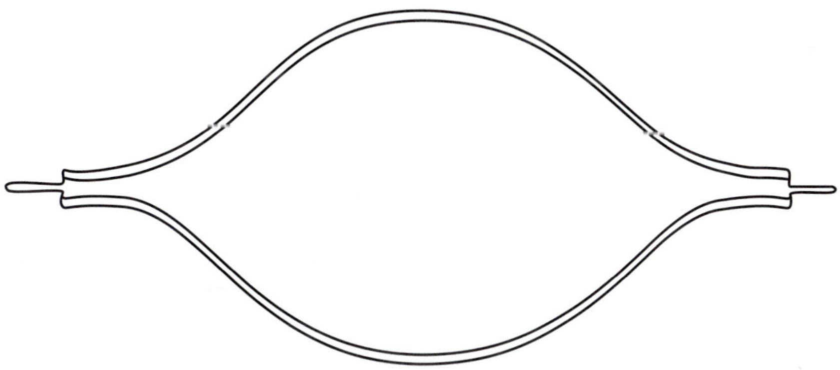

Siphon bulb

   a. What do you think is inside each bulb? Draw a cross-section of both the siphon bulb and the pressure bulb, as shown above. Complete the cross-section by drawing and labeling what you think is inside each bulb.

   b. On your drawings of the bulb cross-sections, add arrows showing which way water flows inside the bulb.

   c. What made one bulb work better than the other?

2. Which type of bulb is better at pumping water? Why?

3. Your heart pumps blood around your body. Would you expect it to work more like the siphon bulb or the pressure bulb? Explain.

22 The Heart: A Muscle

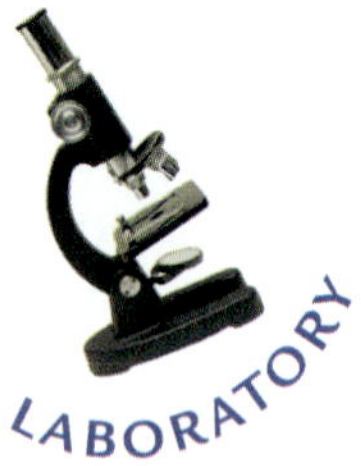

Did you know that your heart is a **muscle?** It's a muscle that works constantly—every minute of every day of every month of every year—for your whole life. You never feel it getting tired, but how hard is it working? Did you know that your heart may beat up to 3 billion times during your lifetime? In this activity you will explore how hard your heart works.

CHALLENGE

How can you measure how hard your heart muscle works?

MATERIALS

For each group of four students

- 1 pressure bulb
- 2 45-cm (18-inch) lengths of clear plastic tubing
- 2 plastic bins (or buckets)
- 1 1-liter bottle or other measuring container
- supply of water
- 1 meter stick (optional)
- paper towels
- access to a wall clock or watch with a second hand

SAFETY

During this activity, water may spill on the floor, so walk carefully.

PROCEDURE

1. Measure and record your resting pulse (in beats per minute). **Hint:** For help on how to do this, review Activity 19, "Heart-ily Fit."
2. Fill one of the bins or buckets ¾ full of water. (Your teacher may have already done this.) The other bin should remain empty.
3. Connect two pieces of plastic tubing to the pressure bulb, one on each end. Place one end of the tubing in each bin.
4. Have members of your group help you keep time and hold the tubing in place as you do the following:
   - **a.** Practice using one hand to pump water from one bin to the other. When you start, be sure to hold the bulb at the same level as the water.
   - **b.** Pour all the water back into one bin so that one bin is again ¾ full of water and the other bin is again empty.

Figure 1: Heart Output	
Height (feet and inches)	**Volume of Blood** (liters per minutes)
4'0"	3.00
4'1"	3.13
4'2"	3.25
4'3"	3.38
4'4"	3.50
4'5"	3.63
4'6"	3.70
4'7"	3.75
4'8"	3.88
4'9"	4.00
4'10"	4.13
4'11"	4.20
5'0"	4.25
5'1"	4.38
5'2"	4.50
5'3"	4.63
5'4"	4.75
5'5"	4.88
5'6"	4.95
5'7"	5.00
5'8"	5.13
5'9"	5.25
5'10"	5.38
5'11"	5.45
6'0"	5.50

c. Use one hand to try to pump the bulb the same number of times as your resting pulse in 1 minute. For example, if your resting pulse is 76 beats per minute, try to pump your bulb 76 times in 1 minute.

d. Record the number of times you were able to pump the bulb in one minute.

e. Use the measuring container to measure the amount of water in the second (initially empty) bin. This is the **volume** you pumped. Record this value in your science notebook.

f. Record in your science notebook how your hand felt after you finished pumping the water.

5. Have each member of your group repeat Step 4 as you help them keep time and hold the tubing in place.

ANALYSIS

1. Compare the pressure bulb model to what you know about your heart. In what ways do you think the pressure bulb is a good model for your heart? What are the weaknesses of the pressure bulb as a model for the heart?

2. Use Figure 1 to find out how much blood your heart pumps per minute based on your height.

a. Record the volume of blood (in liters) pumped by your heart each minute.

b. Compare the amount of blood your heart pumps each minute to the amount of water you were able to pump: Was it more? Was it less? By how much?

3. Describe how hard your heart works by using quantitative and qualitative data from this activity. **Hint:** Be sure to look at your notes from this activity.

4. Why do you think that exercising regularly decreases your heart rate?

23 Heart Parts

READING

In Activity 21, "Inside A Pump," you observed two bulbs: one contained valves and one did not. The valves prevented the backward flow of water, which helped pump water from one container to the other. The human heart has **valves** that help blood flow in one direction. Where are these valves located? Why are they necessary?

CHALLENGE

How does your heart work as a double pump?

MATERIALS

For each student

1 **Student Sheet 23.1, "Heart Diagram"**

red and blue colored pencils

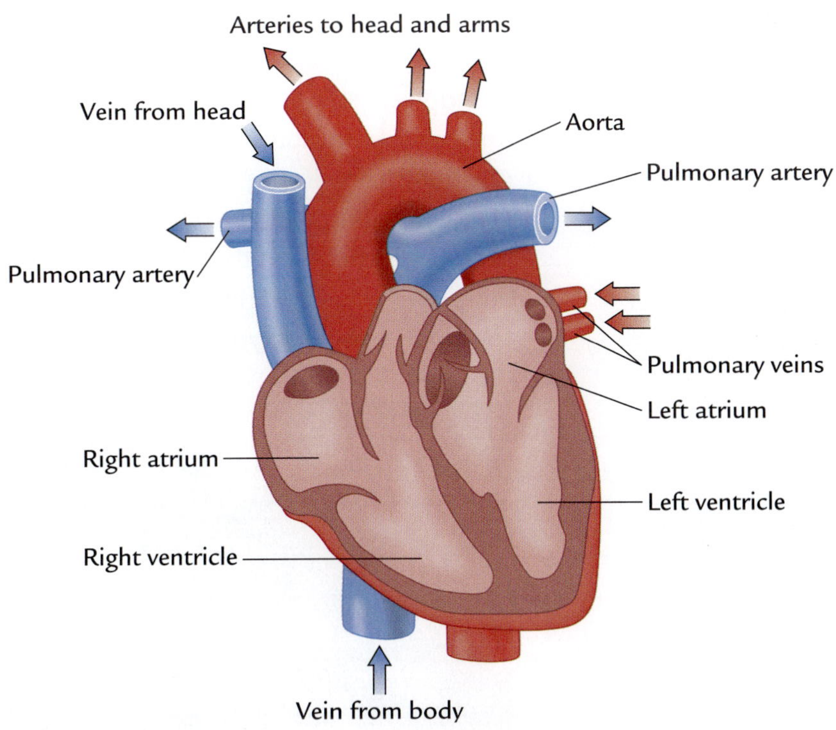

Figure 1: Heart Diagram

READING

Your heart is made up of four chambers that work as two pumps. The right side of your heart acts as a pump that pumps blood to your lungs. The left side of your heart acts as another pump that pumps blood to all other parts of your body. Look at Figure 1 on the previous page. Blood always enters your heart through either of the two chambers known as atria (AY-tree-uh)—the plural for **atrium** (AY-tree-um). Blood is pumped out of your heart through the ventricles—plural for **ventricle** (VEN-trih-kul).

STOPPING TO THINK 1

On Student Sheet 23.1, "Heart Diagram," use your finger to trace the flow of blood through the right side of the heart. Notice where the blood is coming from and where it is going. Repeat this process for the left side of the heart.

There are valves located between your atria and ventricles. There are also valves between your ventricles and the large vessels that lead to the lungs and the rest of your body. The valves open to let blood flow from one structure to another. The valves close to prevent the blood from flowing backward.

When you listen to your heartbeat, you can hear a *lub-dub* sound. The first part of that sound (*lub*) is the sound of the valves between the atria and ventricles closing. The second part of that sound (*dub*) is the sound of a second pair of valves closing. One of these valves is located between the left ventricle and the blood vessel carrying oxygenated blood to the body's organs. The other valve is located between the right ventricle and the blood vessel carrying deoxygenated blood to the lungs.

Medical and scientific illustrators enjoy careers that blend science and art.

STOPPING TO THINK 2

Look carefully at the heart on Student Sheet 23.1. Identify the location of the heart valves. Circle each valve. Then label the valves that produce the *lub* sound and the valves that produce the *dub* sound.

Blood travels through your body in tubes of various sizes. These tubes are known as **blood vessels** (VEH-suls). Blood vessels form a network of tubes throughout your body. At various points in the network, blood vessels are called arteries, veins, or capillaries. Remember finding your pulse in Activity 19, "Heart-ily Fit"? You can feel your pulse as your heart pumps blood through your arteries. **Arteries** (AR-tuh-rees) carry blood away from your heart. Most arteries carry oxygen-rich blood to the organs. The largest artery in your body is the aorta (ay-OR-tuh). **Veins** (VANES) carry blood back to your heart. Most veins carry blood with lower levels of oxygen and higher levels of carbon dioxide (picked up from the organs). Even though the blood in your veins is a deep red, veins look dark purple or blue under the skin of your wrists and under your tongue.

STOPPING TO THINK 3

a. Color your heart diagram on Student Sheet 23.1. Use red for areas that contain blood carrying higher levels of oxygen (and lower levels of carbon dioxide). Use blue for areas that contain blood carrying higher levels of carbon dioxide (and lower levels of oxygen).

b. Recall that arteries carry blood away from your heart and that most arteries carry blood with higher levels of oxygen. Look carefully at the pulmonary (PULL-muh-nair-ee) arteries on Student Sheet 23.1. Explain how these arteries are different from most arteries in your body. **Hint:** Think about the blood they are transporting.

c. Recall that veins carry blood to your heart and that most veins carry blood with lower levels of oxygen. Look carefully at the pulmonary veins on Student Sheet 23.1. Explain how these veins are different from most veins in your body. **Hint:** Think about the blood they are transporting.

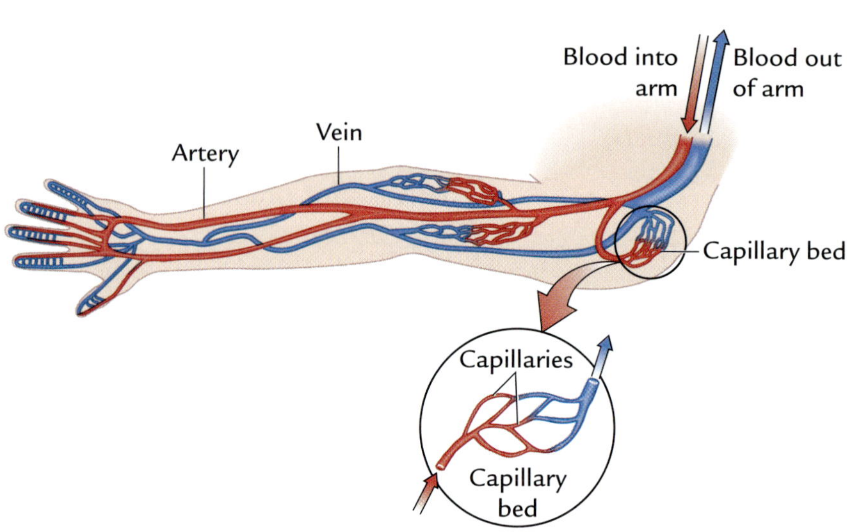

Figure 2: Diagram of Circulatory System (in arm)

Blood leaving the left side of your heart flows into arteries that carry oxygen to all parts of your body. Arteries become smaller and smaller until they become capillaries. **Capillaries** (KA-puh-lair-ees) are blood vessels with walls so thin that oxygen, nutrients, and wastes can pass back and forth. Each capillary is very small and there are many of them. They provide lots of surface area for oxygen to move to the tissues and carbon dioxide wastes to move out of the tissues. As the capillaries widen again, they become veins.

ANALYSIS

1. How is the structure of the heart related to its function?
2. What structures prevent blood in the ventricles from backing up into the atria? Why is it important for your heart to have these structures?
3. Explain what is meant by the statement: "The heart is two pumps." You may want to draw a diagram to support your explanation.

EXTENSION

How did scientists learn about how the human body works? During the 1620s, one scientist, William Harvey, performed dissections to investigate the circulatory system. Research his work to find out more.

24 Round and Round

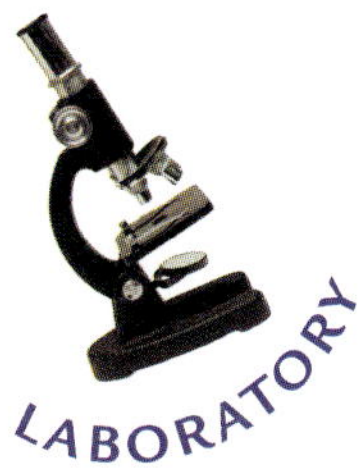

In Activity 18, "The Circulation Game," you modeled the flow of blood through the heart and around the body. In Activity 21, "Inside A Pump," you saw that the heart resembles a pump. Look at the photo below. How well does this model represent the structure of the heart? How well does this model represent the function of the heart?

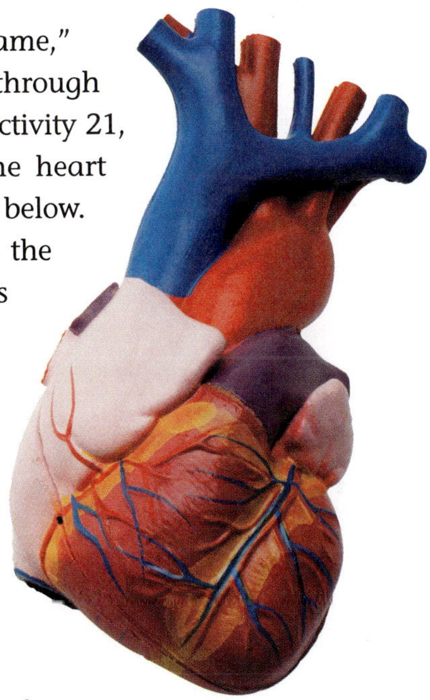

CHALLENGE

How can you construct a model to show how the heart pumps blood to the lungs and the rest of the body?

MATERIALS

For the class

- pressure bulbs
- 45-cm (18-inch) lengths of clear plastic tubing
- plastic bins (or buckets)
- supply of water
- red and blue colored pencils
- siphon bulbs with tubing

PROCEDURE

1. Review the Materials list on the previous page.
2. With your group, design a model using the materials to show how the heart pumps blood to the lungs and the rest of the body. **Hint:** Think about how you modeled the flow of blood through the body in Activity 18, "The Circulation Game."
    - Work together to agree on a design.
    - Listen to and consider the explanations and ideas of other members of your group.
    - If you disagree about a factor with others in your group, explain to the rest of the group why you disagree.
3. Draw a diagram of your model. Label what each part of your model represents.
4. Share your model with another group.
5. Work together to choose one model to test.
6. Set up and test the model.
7. Use your diagram to explain your model to the class.

ANALYSIS

1. How well did your group's original model work? What changes did you make to improve it? Discuss how your design showed how the heart pumps blood to the lungs and the rest of the body.
2. **a.** Draw a diagram of your final model. Use arrows to show which way the water flowed.

    b. Label the parts of your model that represented various organs, structures, or systems of the human body.

    c. Use a *red* colored pencil to identify which tubing contained blood carrying more oxygen and less carbon dioxide. Use a *blue* colored pencil to identify which tubing contained blood carrying less oxygen and more carbon dioxide.
3. Compare the different models that were presented. Which design(s) modeled the function of the heart the best? Explain.
4. **Reflection:** How does modeling help you understand how things work?

25 Healing the Heart

Heart disease is the leading cause of death in the United States. What does that mean? What kinds of problems can your heart have? How have advances in surgery helped people with heart problems?

Dr. Daniel Hale Williams performed the first open-heart surgery in 1893, before x-rays, blood transfusions, and antibiotics. The patient lived for over 50 years after the operation.

CHALLENGE

What are some of the trade-offs involved in developing new treatments for heart problems?

PROCEDURE

1. In order to prepare to watch the story on the video, first read Analysis Questions 1–3.
2. Watch the story of heart surgery on the video, *A Science Odyssey:* "Matters of Life and Death."
3. As you watch, record the advances in surgery that occurred over time. For example, Dwight Harken was one of the first surgeons to perform surgery on a beating heart.
4. Discuss with your class the issues involved in heart disease:
   - What issues regarding experimental design did you observe?
   - Do you have any ethical concerns about what you saw? What are your concerns?
5. Watch the story of organ transplants on the video, *A Science Odyssey:* "Matters of Life and Death."
6. Discuss with your class the issues involved in organ transplants:
   - According to the video, what are the modern issues regarding organ transplants?
   - Which of these issues do you think is the most important? Why?

ANALYSIS

1. What are some of the trade-offs involved in developing new treatments for heart problems?
2. Besides surgery, what are some other approaches to public health problems such as heart disease?
3. **Reflection:** What do you think is the best approach to preventing heart problems for your generation?

EXTENSION

Do you want to know more about the experiments that you saw in the video? Go to the SALI page of the SEPUP website to link to sites with more information on the history of heart surgery.

26 Heart Sounds

INVESTIGATION

How do you know if your heart is healthy? The first thing a doctor usually does is listen to your heart through a stethoscope (STEH-thuh-skope). As you learned in Activity 23, "Heart Parts," your heartbeat sounds like a double beat that is commonly described as *lub-dub*. When blood does not flow normally through the heart, abnormal heart sounds can result.

CHALLENGE

How do heart sounds tell you what is happening inside your circulatory system?

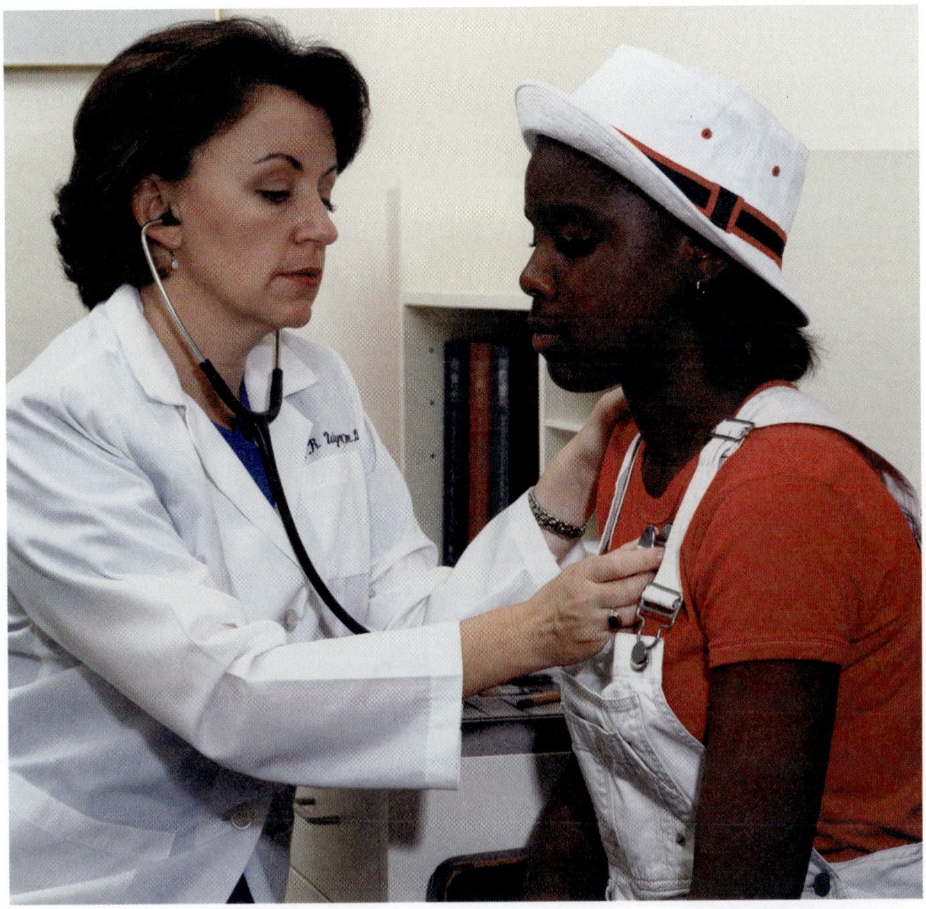

PROCEDURE

Part One: Heart Problems

1. As directed by your teacher, find out more about how the heart works.

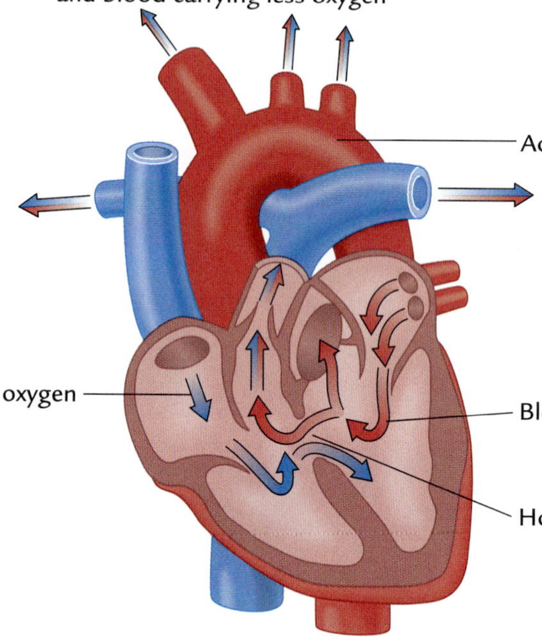

Figure 1: Hole in the Heart Wall

Each person is born with a hole between the left and right side of the heart. This hole usually closes just after birth. But sometimes the hole doesn't close properly. In the past, children born with a hole in the heart would eventually die. Now, with the help of modern surgery, these heart problems can often be repaired.

STOPPING TO THINK 1

Look carefully at Figure 1, "Hole in the Heart Wall." What happens to blood with higher as compared to lower levels of oxygen when there is a hole between the ventricles of the heart?

Think about the function of the valves in the heart pump model. What would have happened if the valves in the pressure bulb didn't work? Some of the water would have flowed backward into the bulb. This would have reduced the forward flow of the water. If heart valves don't work properly, the heart cannot pump blood efficiently and the body does not receive enough oxygen. This can cause breathlessness, exhaustion, and in severe cases, death. Heart valve problems can occur at birth or can be the result of aging or disease.

Compare the normal *lub-dub* heart sound to Figure 2, "Contractions of the Heart." The first sound (*lub*) occurs when the valves between your atria and ventricles close. As your ventricles contract, these valves close to prevent your blood from flowing back into your atria. The second sound (*dub*) occurs when the valves leading to your arteries close and prevent your blood from flowing back into your ventricles.

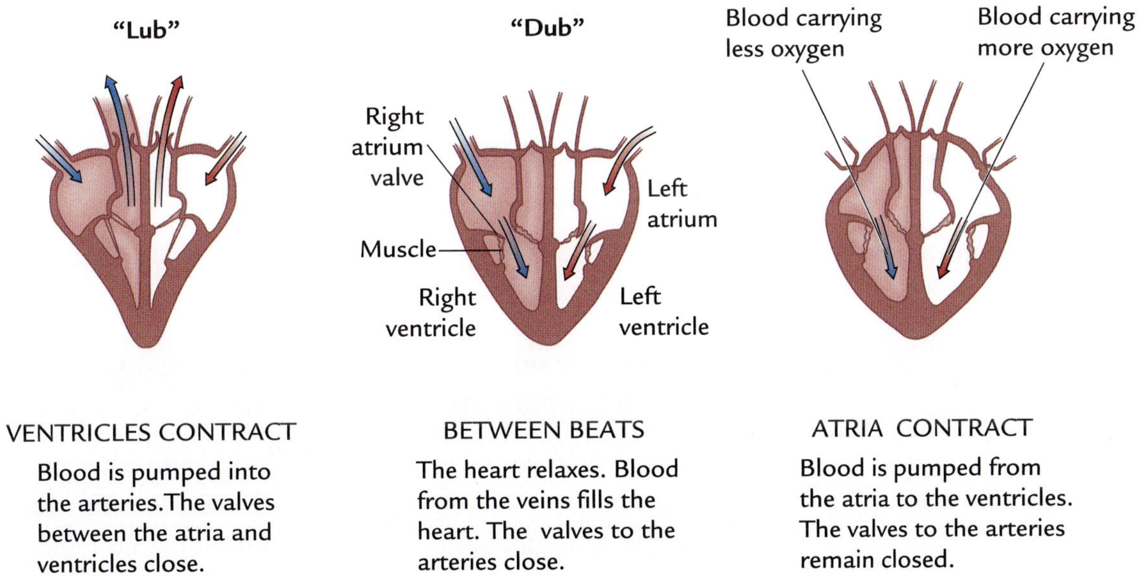

Figure 2: Contractions of the Heart

STOPPING TO THINK 2

Look carefully at Figure 2, "Contractions of the Heart." Discuss the following questions with your group:

- When the left ventricle contracts, where should the blood flow?
- When the left ventricle contracts, the valves between the left atrium and ventricle close. Why? What will happen to blood flow if these valves don't close properly?
- Some heart diseases cause the heart valve openings to narrow. What will happen to blood flow if the valve between the left atrium and ventricle cannot open all the way?
- When the left ventricle relaxes and fills with blood, the valve leading to the aorta closes. What will happen to the blood flow if this valve doesn't close properly?

Part Two: Diagnosis

2. Listen to a series of normal and abnormal heart sounds. As you listen, think about the following questions:
   - Do you hear extra beats in the abnormal heartbeat?
   - Do you hear a whooshing or echo sound?
   - Do you hear an unusual sound during contraction (during and after the first beat)?
   - Do you hear an unusual sound during relaxation (during and after the second beat)?
3. Create a table in your science notebook to record your observations of each heart sound. You will need to record the number of each sound, the kind of sound you hear, and what the heart problem might be.
4. Listen to each abnormal heart sound again. Remember, you will first hear a normal heart sound. This will be followed by the abnormal heart sound. Describe each abnormal heart sound in your table.
5. Think about the types of heart problems that may occur, such as valve problems, a hole in the heart wall, or an irregular heartbeat. Try to identify the type of heart problem that may have caused each abnormal heart sound you heard.

ANALYSIS

1. Cardiologists (kar-dee-AH-luh-jists) are doctors who specialize in the heart. They use heart sounds as one kind of evidence of heart health. What other types of evidence are collected before diagnosis?

2. Why are knowledge of heart valves and the sounds they make important to doctors? Explain.

3. Why do you think the doctor places the stethoscope on several different areas of your chest to listen to your heart beat?

4. Why might someone who has a heart defect such as weak heart valves or a small hole in the heart wall become breathless after climbing a long flight of stairs? Explain.

27 The Pressure's On

LABORATORY

Do you know someone who is concerned about his or her blood pressure? Maybe this person takes medicine or follows a special diet to help control blood pressure. Some people use blood pressure cuffs and stethoscopes to measure their blood pressure. They often keep careful records of their blood pressure from day to day.

What is high blood pressure? How does it affect the ability of your heart to pump blood?

CHALLENGE

What are the effects of high blood pressure on the heart?

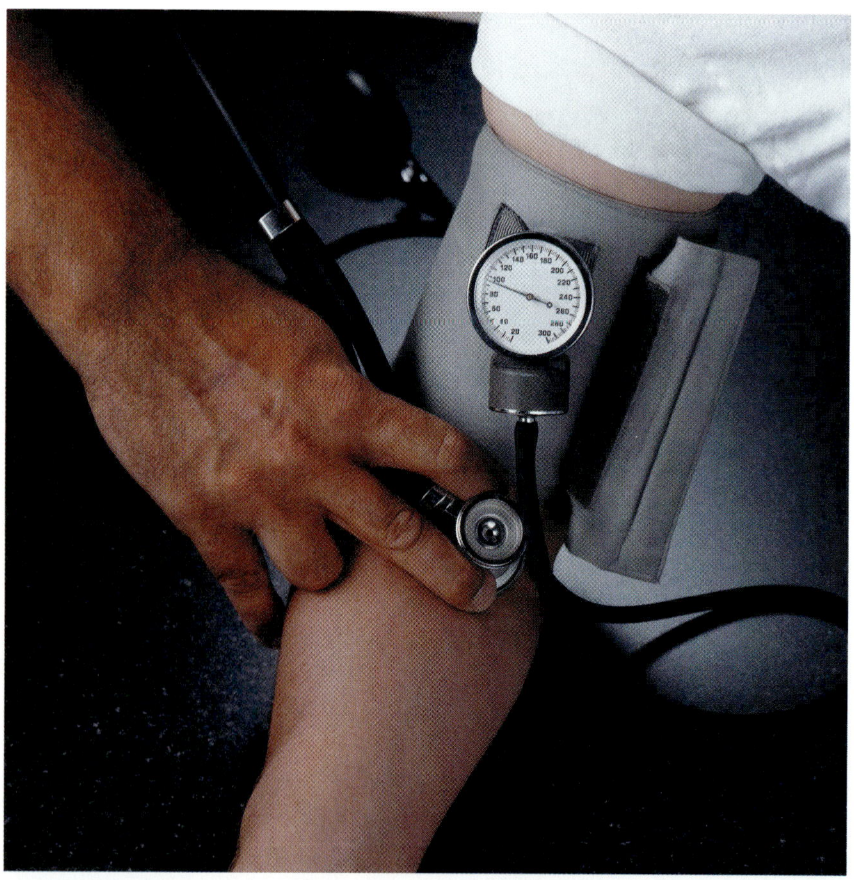

MATERIALS

For each group of four students

1 pressure bulb
2 45-cm (18-inch) lengths of clear plastic tubing
1 plastic bin (or bucket)
1 clamp
1 1-liter bottle or other measuring container
access to a wall clock or watch with a second hand
supply of water

For each student

1 Student Sheet 27.1, "Blood Pressure Data"

PROCEDURE

1. Fill the plastic bin approximately ¾ full of water. Your teacher may have already done this.
2. Place one end of the tubing in the water and establish the flow of water.
3. Each person in your group should perform one of the following roles: (Each member of your group will have a chance to perform each role.)
   a. **Vein:** Hold the tubing in the plastic bin of water. Be sure that the end of the tubing is in the water at all times!
   b. **Artery:** Hold the other end of the tubing in the 1-liter bottle and observe what happens to the water flow at different times while running the model. At the end of the trial, measure the volume of water pumped.
   c. **Timer:** Tell the group when to start and when to stop. Each trial should be 15 seconds long.
   d. **Heart:** Test your ability to pump the water by completing Steps 4–8.

4. In 15 seconds, use one hand to pump as much water as possible from the plastic bin into the bottle.

5. Record the volume of water and your qualitative observations of the effort required to pump the water on Student Sheet 27.1, "Blood Pressure Data." (Use Table 1 for normal blood pressure. Use Table 2 for high blood pressure when using the clamp.)

6. Pour the water from the bottle back into the plastic bin.

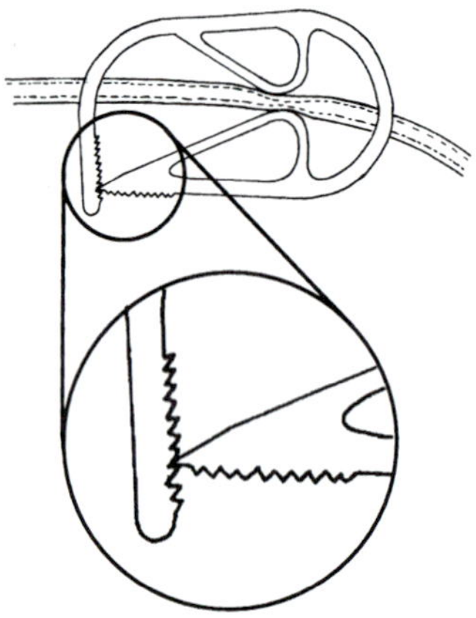

Figure 1: Placing the Clamp

7. Model high blood pressure by placing a clamp on the tubing. Slide the "artery" tubing (tubing leading to the bottle) through the clamp. Tighten the clamp by snapping it down three spaces, or clicks, to the third notch. (See Figure 1.)

8. Repeat Steps 4–6.

9. Rotate roles within your group. Repeat until all members of your group have completed Steps 4–8.

10. On Student Sheet 27.1, calculate the average volume of water the heart was able to pump under normal and high blood pressure.

11. Create a bar graph of the average volume of water the heart was able to pump under normal and high blood pressure. Be sure to label your axes and title your graph.

ANALYSIS

1. Squeezing the pressure bulb modeled the pumping of the heart and the work the heart does when blood pressure is normal.

   a. Recall what you observed when you acted as the artery. What happened to the flow of blood when the artery was clamped?

   b. How did clamping the artery affect how hard the heart had to pump?

2. What effect did clamping the artery have on the pressure of the water inside the tubing? What does it mean to have high blood pressure?

3. Explain what happens to the heart when you have high blood pressure as modeled in this activity. Support your answer with qualitative and quantitative data.

28 Heart Problems

The tissues of your body need a constant supply of oxygen. Remember the circulation model from Activity 18, "The Circulation Game"? Imagine what would have happened if the flow of blood had stopped: the body would not get the oxygen it needs. The body's tissues can survive for only about 4 minutes without a fresh supply of oxygen. Two organs that are especially sensitive to a shortage of oxygen are the brain and the heart itself.

CHALLENGE

What causes a heart attack or a stroke?

MATERIALS

For each student

1 **Student Sheet 28.1, "Problems in the Heart"**

colored pencils (optional)

READING

The Heart Needs Blood Too

Blood goes through your heart as it is pumped. But the blood passing through the chambers of your heart does not supply oxygen to your heart muscle. Some of the blood that leaves your left ventricle travels back to your heart through arteries known as **coronary** (KOR-uh-nair-ee) **arteries**. The coronary arteries lead to a network of capillaries in your heart muscle. These capillaries carry the blood that provides oxygen to your heart muscle.

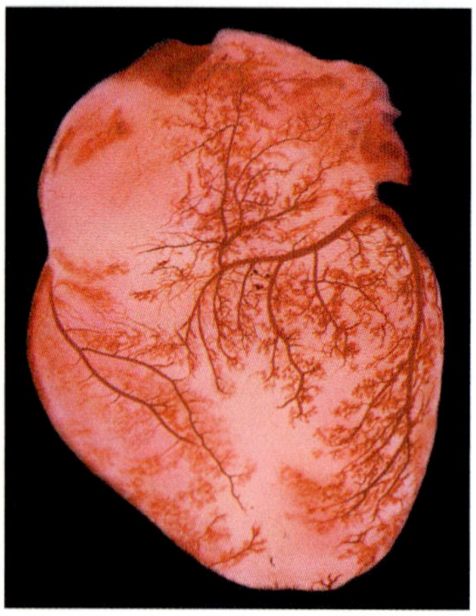

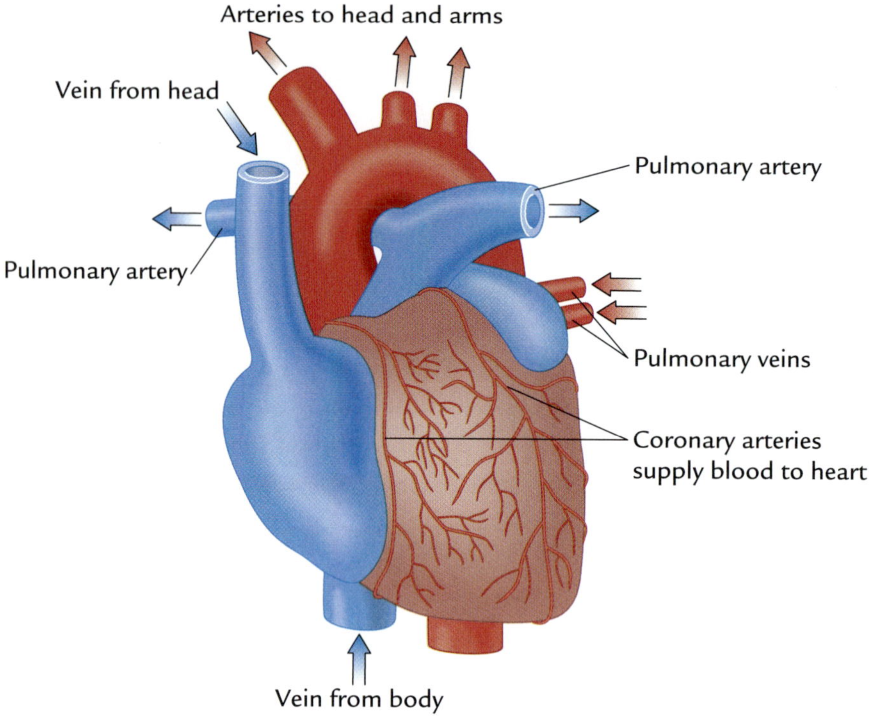

Figure 1: Your Heart

If the blood supply is blocked or reduced, heart muscle can be damaged and pain results. This is commonly known as a heart attack. If a large enough part of the heart muscle is affected, then the heart can no longer pump blood. Heart failure and death result.

STOPPING TO THINK 1

a. Look at Stopping to Think 1 on Student Sheet 28.1, "Problems in the Heart." Think about what would happen if the blood flow were blocked at Point A on the diagram. On Student Sheet 28.1, shade the area of heart muscle that would be affected.

b. Think about what would happen if the blood flow were blocked at Point B on the diagram. On Student Sheet 28.1, shade the area of heart muscle that would be affected.

c. Which heart attack would be most likely to kill a person: the one caused by a blockage at Point A or the one caused by a blockage at Point B? Explain.

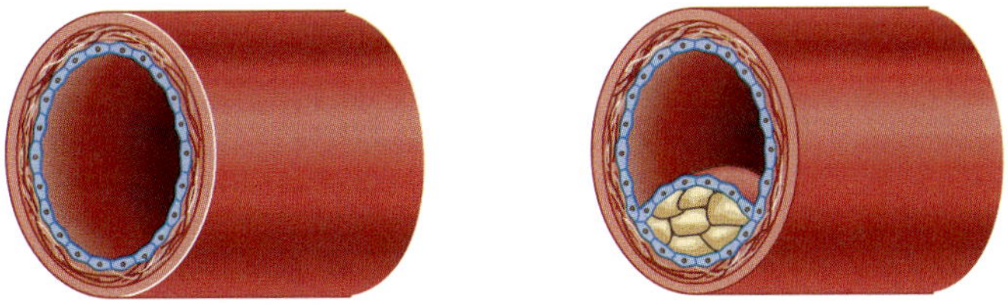

Figure 2: Normal and Blocked Arteries

How Heart Attacks Happen

Arteries supplying blood to the heart can become blocked by blood clots or by fat deposits. Look at Figure 2. In one condition called atherosclerosis (a-thuh-row-skluh-ROW-suss), deposits of fat partially block the center of the arteries. A diet high in some kinds of fat is associated with a higher risk of this condition. People who have high levels of this fat, called cholesterol, in their blood may modify their diet to reduce their cholesterol level.

STOPPING TO THINK 2

a. Look at Stopping to Think 2 on Student Sheet 28.1. In Artery A, draw a fat deposit that would block the artery just a little. In Artery B, draw a fat deposit that would block over half of the blood flowing through the artery.

b. Which of the two arteries (A or B) would be more likely to become completely blocked if a small blood clot came through the artery?

c. Why are people who are recovering from heart attacks sometimes given medicine to reduce the clotting activity of their blood?

Medical Help for Blocked Arteries

There are two common treatments for blocked arteries in the heart. In one treatment called coronary bypass, the surgeon inserts a segment of a vein—removed from another part of the body—around the blocked artery. This lets the blood flow through another pathway, bypassing the blocked artery. In another treatment called angioplasty [AN-jee-uh-plas-tee]), the surgeon inserts a tube with a balloon-like device into the partially blocked artery. The balloon is then inflated to stretch the artery. After the balloon is deflated and removed, the artery will be more open than before.

STOPPING TO THINK 3

Look at Stopping to Think 3 on Student Sheet 28.1. Imagine that tests reveal that a person has a blockage developing in the artery at Point X on Figures 4 and 5. Doctors are planning to do a coronary bypass operation. Does Figure 4 or Figure 5 show a bypass that will help this person's problem?

High Blood Pressure and Heart Disease

As the strongest muscle in your body, your heart first pumps blood through your arteries. This pumping action creates pressure within your blood vessels. As you discovered in Activity 27, "The Pressure's On," high blood pressure forces the heart to work harder to pump blood through the circulatory system. This can cause the heart muscle to become enlarged and weakened. It also strains the arteries, causing them to lose their ability to stretch. They become harder and less elastic. This is known as "hardening of the arteries." This process occurs with age, but it is more common in people with high blood pressure. With damaged arteries, blood clots or deposits of fat are more likely to form. When the arteries that supply the heart are damaged, a heart attack can follow. (A similar type of attack, known as a stroke, can happen if the arteries to the brain are damaged.)

In most cases, researchers do not know exactly what causes high blood pressure. Several factors, though, seem to increase a person's chance of having high blood pressure. Factors that increase the chance of an illness are called **risk factors**. Risk factors for high blood pressure include the person's age, weight, race, gender, and body type; whether the person smokes cigarettes; how much alcohol the person drinks; the person's diet and amount and frequency of exercise; and heredity (whether the person's parents or grandparents have high blood pressure). Older people, people with a family history of high blood pressure, people of African descent, and men in general have a higher risk of high blood pressure and other types of heart disease than other groups. Being extremely overweight, living an inactive life style, drinking large amounts of alcohol, and, in some people, eating a diet high in salt are associated with a greater risk of high blood pressure.

Smoking and Heart Disease

Cigarette smoking increases the risk of death from heart disease because of its impact on several risk factors. Smoking can increase blood pressure, increase the formation of fat deposits in the arteries, and increase the chance of blood clot formation in the arteries. A smoker is also less likely to survive a heart attack. The smoker's lungs are not as efficient in absorbing oxygen from the air and thus provide less oxygen to maintain the heart.

ANALYSIS

1. What kinds of health problems can be caused by blockages in coronary arteries?
2. Why should people with many risk factors for heart disease first check with a doctor before beginning an exercise program?
3. **Reflection:** What can you do to maintain or improve the health of your heart?

29 Helping Hearts

PROJECT

Having healthy habits now can reduce your risk of heart disease later. What factors increase your risk of heart disease? Which of these risks can you reduce? Find out by identifying which heart disease risks are **voluntary**—that is, under your control.

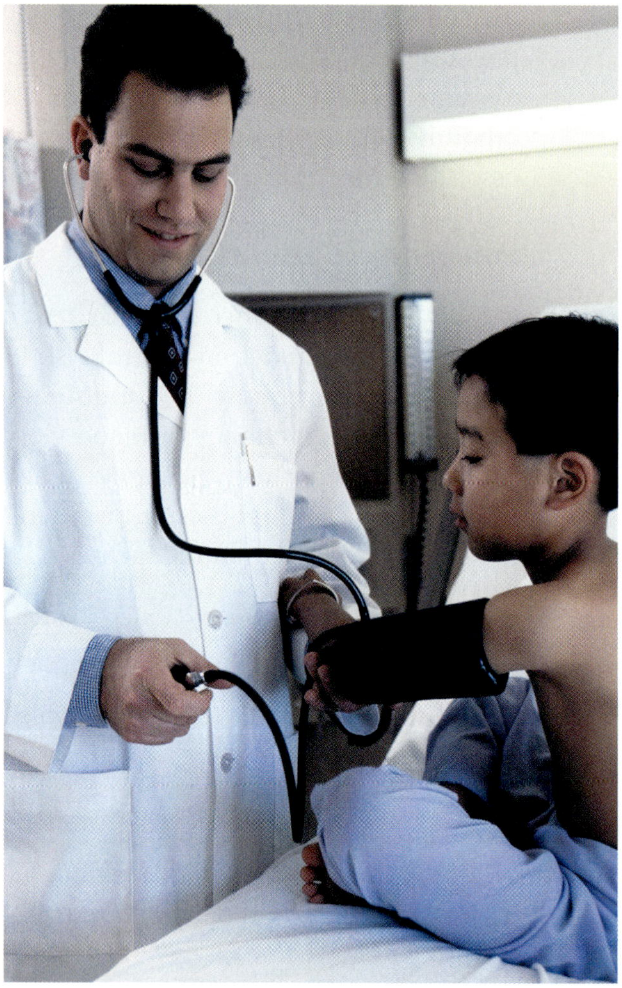

Regular check-ups help people find out if they have high blood pressure or other risks for heart or circulatory disease.

CHALLENGE

What is your relative risk of heart disease? How can you convince people to make choices that reduce their level of risk?

MATERIALS

For each student

1 **Student Sheet 29.1, "Heart Risk Quiz"**

PROCEDURE

Part One: Heart Risk Quiz

1. Evaluate your personal risk for heart disease by taking the quiz on Student Sheet 29.1, "Heart Risk Quiz." This quiz should be completed at home so you can ask questions of your family members. You will not have to share the results in class. Use the Scoring Guide on the back of the sheet to add up your risk points.

2. Look again at Student Sheet 29.1, "Heart Risk Quiz." Note that each item on the quiz is either about a risk (such as eating high-fat meals) or about a way to reduce a risk (such as eating fruits, beans, and vegetables).

Decide with your group which risks are voluntary. If you think a risk is only partially voluntary, discuss why. Try to come to agreement within your group.

- Listen to and consider the explanations and ideas of other members of your group.
- If you disagree about a risk with others in your group, explain to the rest of the group why you disagree.

3. Many people know the risks for heart disease, but not all people work hard to reduce their risk. Look at the voluntary risks you identified in Step 2. Discuss with your group the possible reasons why young people like yourself may not change their habits.

Part Two: Helping Others Make Healthy Decisions

4. Design and produce a brochure persuading teenagers to make decisions to reduce their risk for heart disease.

   Your brochure should do the following:

   - Be clear and convincing.
   - Describe one voluntary risk factor for heart disease.
   - Explain how the circulatory system works.
   - Explain how the risk factor you have identified affects the structure and function of the circulatory system.
   - Suggest practical steps teens can take in their daily lives to reduce this risk factor.
   - Use informal diagrams and artwork to help present the content of the brochure.

ANALYSIS

1. Imagine you are a doctor and Mr. Jacobs visits you to discuss his health. He says, "I feel great. I run 2 miles every day, lift weights, and eat a healthy diet. No red meat and fast food for me! I looked at the risk factors for heart disease and it looks like I have zero risk. I'm sure I'll never have any problems!"

   What would you tell Mr. Jacobs about his risk for heart disease? Support your answer with evidence.

2. Imagine you are a doctor and Ms. McDonald visits you to discuss her health. She says, "I just took a look at risk factors for heart disease. I fall into a lot of high-risk categories. I work long hours and don't have much spare time. I eat a lot of fast food hamburgers and fries, and I exercise only on Saturdays. Does this mean that I will definitely have heart disease?"

   What would you tell Ms. McDonald about her risk for heart disease? Support your answer with evidence and discuss any trade-offs involved in your recommendation.

Hint: To write a complete answer, first state your opinion. Provide two or more pieces of evidence that support your opinion. Then consider all sides of the issue and identify the trade-offs of your decision.

EXTENSION

Visit the SALI page of the SEPUP website for links to more information about heart disease risk. Do you think the materials you find there are successful at convincing readers to reduce their risk?

Index

A

B

C

D

E

F

G

H

I

K

L

M

N

O

P

Q

R

S

T

U

V

W

Credits

Abbreviations: t (top), m (middle), b (bottom), l (left), r (right), c (center)

All illustrations by Seventeenth Street Studios, except as follows: Pages B-26, B-36, B-43, B-63, B-66, B-72, B-73, B-80, B-81: Precision Graphics.

"Talking It Over" icon photo: ©Michael Keller/The Stock Market

Unit A

Unit opener (A-2, A-3): bl: ©S. Fraser/Photo Researchers, Inc.; bc: ©2001 Richard Price/FPG; br: Donna Markey; tl: ©2001 Charles Thatcher/Stone; tc: ©2001 B. Busco/The Image Bank.

A-8 ©2001/The Image Bank; A-9 ©Bettmann/CORBIS; A-25 ©Digital Art/CORBIS; A-34 Heart Pacemaker ©Department of Clinical Radiology, Salisbury District Hospital/SPL/Photo Researchers, Inc.; A-44 photo courtesy of the World Health Organization; A-46 ©Laura Dwight/DoctorStock.com.

Unit B

Unit opener (B-3): br: Donna Markey.

B-37 (photo) ©Martin Dohrn/Royal College of Surgeons/SPL/Photo Researchers, Inc.; B-69 courtesy of the Provident Foundation, Chicago, Illinois; B-76 (photo) ©SPL/Photo Researchers, Inc.

Cover photo (kids running): ©2001 David Young-Wolff/Stone